ALTERNATIVE IRRIGATION

ALTERNATIVE IRRIGATION
THE PROMISE OF RUNOFF AGRICULTURE

Christopher J Barrow

Earthscan Publications Ltd, London

First published in the UK in 1999 by
Earthscan Publications Ltd

A catalogue record for this book is available from the British Library

ISBN: 1 85383 496 3 paperback
          1 85383 495 5 hardback

Typesetting by PCS Mapping & DTP, Newcastle upon Tyne
Printed and bound by Creative Print and Design Wales, Ebbw Vale
Cover design by Yvonne Booth
Cover photograph © Christopher J Barrow
    *Valleyside irrigated terraces, supplied by channel (gravity flow) from meltwater
    stream, Azilal province, High Atlas Mountains, Morocco, 1998*

For a full list of publications please contact:

Earthscan Publications Ltd
120 Pentonville Road
London, N1 9JN, UK
Tel: +44 (0)171 278 0433
Fax: +44 (0)171 278 1142
Email: earthinfo@earthscan.co.uk
http://www.earthscan.co.uk

Earthscan is an editorially independent subsidiary of Kogan Page Limited and
publishes in association with WWF-UK and the International Institute for Environment
and Development

This book is printed on 50 per cent recycled paper

CONTENTS

LIST OF FIGURES, BOXES AND TABLES

FIGURES

BOXES

TABLES

LIST OF PHOTOGRAPHS

LIST OF ACRONYMS AND ABBREVIATIONS

BOD	biological oxygen demand
CIS	Commonwealth of Independent States
CO_2	carbon dioxide
EEC	European Economic Community
ENSO	El Niño Southern Oscillation
EU	European Union
FAO	Food and Agriculture Organization
FELDA	Federal Land Development Authority
FSA	farming systems analysis
FSR	farming systems research
FSR&E	farming systems research and extension
GIS	geographical information system
ha	hectare
ICRISAT	International Centre for Research in the Semi-Arid Tropics
IFAD	International Fund for Agricultural Development
IIED	International Institute for Environment and Development
ILRI	International Livestock Research Institute
m	metre
m^2	square metre
m^3	cubic metre
mm	millimetre
NaCl	sodium
NGO	non-governmental organization
ODI	Overseas Development Institute
PRA	participatory rural appraisal
RRA	rapid rural appraisal
RUSLE	revised universal soil-loss equation
SADCC	Southern African Development Co-Ordination Conference
SAREC	Swedish Agency for Research and Cooperation with Developing Countries
SIDA	Swedish International Development Agency
SWC	soil and water conservation
UNEP	United Nations Environment Programme
UNESCO	United Nations Educational, Scientific and Cultural Organization
USAID	US Agency for International Development
USDA	US Department of Agriculture

USLE	universal soil-loss equation
USSCS	US Soil Conservation Service
USSR	Union of Soviet Socialist Republics
UV	ultraviolet
WEDC	Water, Engineering and Development Centre (Loughborough University, UK)
WSC	water and soil conservation
WT	water-table
yr	year

PREFACE

Runoff agriculture makes use of surface and subsurface flows of water, such as stormwater, ephemeral streamflow and floodwater. When successful it can significantly improve upon rain-fed agriculture, offering one or all of the following: improved security of harvest, increased yields, sustainable production, and opportunities for crop diversification. These benefits can often be gained without some of the unwanted environmental and socio-economic impacts frequently associated with other approaches to upgrading rain-fed agriculture, especially mainstream irrigation methods.

Runoff agriculture offers opportunities that small-scale farmers and pastoralists would probably not otherwise be able to afford. In some regions rain-fed cultivation and pastoralism are in a state of decline and something is needed to counter the loss of livelihood and frequently associated environmental degradation. Runoff agriculture and the related approaches covered in this book have particular value for remote and harsh environments where other strategies would be either technically impossible, too expensive or ill-advised. In addition, runoff agriculture often uses locally available materials and hand labour, so it can be adopted with little risk of dependence, may spread with limited outside aid and is adaptive. With food production in many developing countries falling, land degradation increasing and uncertainty about future climate changes as a consequence of global warming, runoff agriculture will become much more important.

Runoff agriculture had had much less support than mainstream irrigation, in terms of research expenditure, investment in implementation and in extension services. Commercial interests will probably continue to neglect approaches such as runoff agriculture, which do not lead to large sales of inputs, and will continue to invest in mainstream irrigation. However, non-governmental organizations, governments and aid agencies could do much more to promote and support alternatives such as runoff agriculture and getting them to do this seems a realistic goal.

Wherever runoff agriculture is supported it is important that those involved are aware that, although runoff agriculture has potential, it can go wrong like mainstream irrigation. Small-scale schemes and the participation of local people do not guarantee success; careful choice of approach and sensitive implementation of whatever works well in a given situation, plus effective research and extension services are required.

Christopher J Barrow
University of Wales Swansea
October 1998

Acknowledgements

I wish to thank Hamid Hicham and staff of the Direction Provincial d'Agriculture d'Azilal (DPA), based in or around Azilal in Morocco for facilitating studies of traditional terrace agriculture. My thanks are also due to The British Academy for providing field visit funding. I am also most grateful to the Library, University of Wales Swansea.

1 INTRODUCTION

It is clear that a few years from the start of the 21st century, the world's agriculture, in spite of impressive achievements over the last century, faces serious challenges (see Box 1.1). Population growth is one of these – between AD 1900 and 1992 the world's human population grew from roughly 1.6 billion to over 5.0 billion, and it is unlikely to stabilize before reaching 11 billion and may reach 15 billion by AD 2050. Irrigated land increased roughly five-fold to around 235 million hectares between 1900 and 1992, so that by the early 1990s about 16 per cent of total cropland was 'irrigated' and gave approximately 36 per cent of total harvest. More than half of increased food production since 1970 has come from irrigated land (rough estimate) (Postel, 1992, p49; Srivastava et al, 1993, p13). Irrigation is particularly important in Asia where it provides about two-thirds of all food supplies although it comprises less than half of the cultivated land area (Le Moigne et al, 1989, pvii). For some areas of the world, increasing agricultural output through irrigation supplied by river diversion or groundwater will be impractical; however, agricultural improvement may well be possible through better runoff management. Roughly 84 per cent of the world's agricultural land is rain-fed and either needs improvement or has potential for improvement.

It is important to find alternatives to mainstream irrigation development (mainstream implies large scale and high tech), not only because many localities are unsuitable for it and because of growing competition for available freshwater supplies, but also to reduce pollution and other impacts often associated with established irrigation approaches. These impacts should not be underestimated; there are too many examples of large numbers of people who have suffered socio-economic and health problems, of land degraded and biodiversity destroyed. For example, the Aral Sea bears testimony to ill-conceived irrigation development. In 1961 the Aral Sea was the world's fourth largest inland waterbody, with rich fisheries and flourishing communities around it; by 1990 its volume had shrunk by 69 per cent and the water was badly contaminated by salts, pesticides and fertilizer from the return flows of cotton and rice irrigation in the southern Commonwealth of Independent States (CIS, former USSR). Today the fisheries have virtually disappeared, settlements have withered, wildlife has suffered badly, and many people are ill from

---

Box 1.1 CHALLENGES FACED BY AGRICULTURE IN THE LATE 20TH CENTURY

Challenges include:

* growing population;
* soil degradation;
* breakdown of established social controls and obligations affecting agriculture practices which leads to environmental degradation and rural–urban migration;
* growing competition for water supplies;
* risk of global environmental change;
* pollution of various kinds affecting agriculture (for instance, acid deposition);
* risk of further stratospheric ozone depletion raising UV-radiation damage;
* failure to invest enough in improving the agriculture of poor smallfarmers, especially those in harsh (marginal) environments;
* loss of access to common resources;
* rising costs of establishing new large scale agriculture (especially irrigation);
* low market prices and marketing difficulties for smallfarmers;
* escalating costs of rehabilitating failed agricultural development;
* outmigration from rural areas causing labour shortages;
* environmentaly harmful agrochemicals (pesticides, herbicides, artificial fertilizers).

---

the effects of pollution. Worse, the irrigation that has caused these impacts may not be sustainable in the long-term or even reasonably profitable in the short-term (Adams, 1992; Jones, 1997, p217; Kobori and Glantz, 1998).

Although mainstream irrigation has contributed a great deal since the 1950s to feeding people and providing agricultural commodities, this must be weighed against the fact that, since the turn of the century, a large proportion of expenditure on agricultural research and much of the total investment in agriculture has been directed towards it. At the time of writing, over 60 per cent of all agrochemicals were used by the irrigation sector, which includes most larger-scale producers. The success of irrigation is thus hardly surprising and begs the question: what would similar investment yield if directed at runoff or rain-fed agriculture and smallfarmers?

The reality is that the runoff and rain-fed agriculture sectors are neglected and, unless attitudes change, efforts to upgrade (to intensify) agriculture will be based on expensive mainstream irrigation which is often poorly adapted to local environments and the needs of many rural people (notably the production of locally available food and the generation of employment) and is seldom sustainable (Tully, 1990). It is not cost effective to extend mainstream irrigation to some parts of the world. For example, it has been estimated that only 5 per cent of sub-Saharan agricultural land is under mainstream irrigation, and with costs rising fast, cheaper and more accessible alternative ways of boosting yields, improving security of harvest and sustaining agriculture are needed.

Total water use (for example, for agriculture, domestic supply and industry) has increased five-fold since AD 1900 (Uitto and Schneider, 1997, p5). In preparation for the Rio Earth Summit of 1992, the International Conference on Water and Environment met in Dublin to discuss water availability and its constraints. Irrigation plays a very important role in world food production and uses roughly 65 per cent of total available freshwater; that use is projected to fall to roughly 60 per cent by AD 2000 as demand for domestic supplies and industry increases (Agnew and Anderson, 1992; Pearce, 1992; Hillel, 1994, p34). As water availability is falling, world population is increasing: by AD 2040 there will probably be over nine billion people to feed (compared with the present estimated 5.5 billion) and less water for irrigation. According to Pereira et al (1996) world food production must double in the next few decades. Postel (1992, p58) warned: 'With some 95 million people being added to the planet each year in the nineties, new strategies will be needed to prevent the many irrigation constraints from leading to food shortages.' The struggle to find enough water for cities and to feed growing populations is already a threat to well-being for some regions and countries.

Large irrigation schemes often leak water from their supply channels and crop land; it is not unusual for 80 per cent of the water diverted from a river or pumped from groundwater to fail to be absorbed by crops (Barrow, 1987, p209). More efficient agricultural use of water and alternative sources must be urgently pursued. Before the late 1970s irrigation expanded fast enough to keep pace with population growth, but recently there has been a per capita decline in irrigated land area and food security is falling as a consequence (Postel, 1992, p51). Governments and development agencies have mainly supported large scale, commercially oriented mainstream irrigation develop-ment projects which rely on considerable quantities of water diverted from rivers or groundwater and suitable land. Both these resources are becoming more difficult to find. Given the commercial support and government favour it enjoys, mainstream irrigation will continue to be implemented. At least there are signs that the mainstream irrigation sector is paying more attention to rehabilitating degraded schemes and to developing water conservation technology, especially low-cost (but still too expensive for many agricultural-ists), low-volume drip, bubble and trickle water application techniques (Barrow, 1987; Lambert and Faulkner, 1989; Le Moigne et al, 1989). As well as improving mainstream irrigation it is important to develop and promote acces-sible, sustainable , effective, and water-conserving alternatives, and runoff agriculture offers some of the most promising strategies for doing that.

In rich and poor nations irrigation development has too often resulted in unwanted consequences: groundwater supplies have been depleted and sometimes fail permanently; agrochemical-charged return flows contaminate groundwater, lakes, rivers and seas (such as the Aral Sea); wetlands are damaged and coral reefs may suffer; soils may be salinized and ruined; debts are incurred; people may be relocated without adequate compensation; and water-related illnesses may increase. Many large irrigation schemes fail to adequately repay investment and often degenerate, sometimes necessitating abandonment so that much agriculturally productive land is lost. Where large scale irrigation produces export crops or foodgrains for urban areas, small farmers may find that market prices fall. They then neglect or abandon produc-tion to the detriment of their livelihoods and national food security.

Low rewards for smallholder agricultural production can prompt rural folk to migrate to cities, where they usually join the ranks of the unemployed. Thus, the benefits of investment in large-scale commercial irrigation may be overshadowed and the impacts may marginalize other off-site agriculturalists, enough to cause urban migration and widespread environmental degradation. For example, in Morocco the government has invested in large irrigation schemes mainly in the north-western lowlands. The impact of the cheap crops these produce, including grain purchased from abroad with the profits of export crops, together with the lack of investment in traditional farming, has led to overgrazing of common land, narcotics cultivation and neglect of traditional arable farming in the Rif and Atlas Mountains. Consequent land degradation has helped to silt up dams and channels and has reduced groundwater recharge, threatening some of the large scale lowland irrigation schemes and urban water supplies. These are high prices to pay for irrigation development that generally offers limited employment generation and often vulnerable and unsustainable production and increased dependency (for instance on pump parts, agrochemicals and fuel).

The problems associated with large scale irrigation are well known and funding agencies and irrigation experts call for better engineering and management and for rehabilitation of schemes which have failed (perhaps as much as two-thirds of the world total). However, as easier-access water supplies and the best land are used, costs of large scale irrigation spiral up; by 1992 a big scheme in Africa often exceeded US\$ 20,000 per hectare to establish (Postel, 1992, p52). In spite of the costs and risks nations still plan to develop large irrigation; for example, Egypt has been reviewing proposals to divert up to 10 per cent of Nile flows to irrigate the Western Desert. The expense would probably be well over UK£1.2 billion (at 1997 prices) and the development could be environmentally and socially damaging and unsustainable (Pearce, 1997, p5). Commercial interests still see potential for profit if they invest in mainstream irrigation – there are pumps, water distribution equipment and agrochemicals to be sold; runoff agriculture does not offer such opportunities for gain and so will probably have to be supported by non-commercial means.

The opportunities for opening up new land for rain-fed agriculture are decreasing – although the development of drought- and salt-tolerant crops may help, and there are regions of Africa where underpopulation and communal landholding may prevent the intensification of agriculture (Tiffen et al, 1994). Outside the humid tropics, and even within some humid tropical localities, rain-fed agriculture runs the risk of 'uncertain rain' (Adams, 1992); to improve security of harvest (as well as to boost crops and perhaps allow diversification) some way of improving moisture availability is required. In many parts of the world that will have to be through runoff agriculture because there are no suitable groundwater or river irrigation water supplies, or they are too distant, or some company or city has appropriated them.

Growing human populations are more likely to be fed through *intensification* rather than *expansion* of farming. As already discussed, intensification in recent decades has mainly occurred through promoting large scale irrigation. However, the number of water-scarce regions and countries is growing:

Postel (1992, p29) estimated that at least 232 million people were already affected. Some countries are relying on finite groundwater supplies and have few or no rivers as an irrigation supply alternative when underground supplies run out; for example, Libya has spent huge sums on piping groundwater to large irrigation schemes from sources that may fail within 50 years. Shortage of water supplies for irrigation not only threatens world food supplies, it is, for some regions, a serious threat to maintaining peace (Hillel, 1994; Ohlsson, 1995). The possibility of water shortage-related conflicts may increase if global climate change affects supplies and demands (Parrey, 1991).

Worldwide, topsoil is being lost or degraded at a very worrying rate; during the last 40 years it has been suggested that nearly one third of the world's arable land has been damaged by erosion and estimates have suggested the loss continues at a rate of more than ten million hectares a year (Pimentel et al, 1995). In addition to losses to erosion, land is also lost to production through declining fertility and salinization). Land used for crops, grazing and forestry is also suffering degradation from salinization, acidification, loss of organic matter due to poor landuse and brushfires, accumulation of pollutants, and unwanted soil structure changes. Soil degradation seriously threatens efforts to raise food and commodity crop production in developed and developing countries (the Food and Agriculture Organization – FAO – has reported falls in per capita food production in recent years as a consequence of soil degradation). So far, even in developed countries, disappointing progress has been made in countering soil degradation. It is also important not to allow the continued decline in food production by smallfarmers and to sustain pastures and forests.

It is quite common to find that rural livelihood strategies have broken down or are probably about to. The reasons are manifold, and include: population increase; globalization; penetration of capitalism into subsistence economies; social change; civil unrest; and the impact of structural adjustment programmes, to name but a few. The challenge is to develop runoff agriculture to counter such breakdown and to limit the resulting misery and environmental damage it would cause.

RUNOFF AGRICULTURE AND LAND HUSBANDRY

It is vital not to separate agricultural improvement and soil and water conservation (SWC) measures, and runoff agriculture does link them. Often agriculture suffers from both poor or declining soil fertility and shortage of moisture. Sustainable and improved-yield agriculture (or establishing vegetation cover or forestry) demands care of soil and water resources, especially in harsh environments. This is best achieved through land husbandry. This is more than efficient management: it is at least stewardship of resources and, ideally, their improvement while sustaining harvests. It is also something that should be achieved without unwanted impacts on surrounding or distant areas, cruelty to livestock, loss of biodiversity, or exploitation of agricultural employees. Hudson (1992, p9) made a plea for 'a positive approach where care and improvement of the land resource comes first, and control of erosion follows as a result of good *land husbandry*' [Hudson's italics]. A similar concern has been voiced by Adams (1992, p16): 'what is needed is a new approach to development, combining integrated natural resource management with realistic socio-economic goals'. These pleas are typical of recent calls for holistic land husbandry approaches to ensure proposed agricultural improvements will continue to work in given (but possibly not static and unchanging) environmental and socio-economic circumstances (Richards, 1985, p53; Shaxson et al, 1997). Better land husbandry is an important goal that might be achieved via improved SWC and runoff agriculture.

For agricultural development to be sustainable it is not enough that it functions satisfactorily in terms of economics or engineering – it must also 'fit' socially and environmentally, satisfy local and, where necessary, wider market needs and be adaptive to future changes (Brklacich et al, 1991; Altieri, 1995). Good land husbandry combines sustainability with profitability (Hudson et al, 1993, p225), and to get it requires cooperation between, and overall coordination of, farmers, government bodies, non-governmental organizations (NGOs), research bodies, extension agencies, and funding agencies. To sustain agriculture demands, at the very least, adequate moisture, maintenance of soil fertility, avoidance of salinization, prevention of excessive erosion, and control of pests and diseases. Runoff agriculture can do these things in localities where rain-fed agriculture is either not viable or is insufficiently productive or insecure, and where mainstream irrigation would be difficult or uneconomic. The distinction between rain-fed and runoff agriculture is often blurred, as is the difference between flood and wetland agriculture and lift or pump irrigation; often more than one strategy is practiced together.

In addition to runoff agriculture there are other alternatives to today's large scale irrigation approach to agricultural intensification: desalination and low-waste irrigation techniques; development of salt-tolerant crops; more efficient irrigation; use of waste water – all of these have great potential. However, there is already a rich tradition of SWC and runoff agriculture, and some of these approaches or improved versions are appropriate for adoption by the huge numbers of small-farmers practising rain-fed farming, and by pastoralists and those involved in conservation, often in remote and rain-deficient environments which cannot support irrigation and where funding is hard to come by (Hudson, 1987). A number of factors push or pull people into agriculturally marginal areas, and once there it is difficult for them to sustain satisfactory livelihoods and it may become even more of a challenge if global climate change occurs (Glantz, 1994; Valdez et al, 1994). Runoff agriculture offers ways of meeting these challenges.

It is difficult to sustain adequate livelihoods in marginal and sensitive lands which rely on rain-fed agriculture without threat of hardship, hunger and environmental damage, especially when the population is increasing and in the face of likely global environmental change. There are two obvious responses: draw populations away from sensitive areas; improve agricultural strategies in the affected areas, reducing dependence on rain-fed agriculture. The first makes sense only if people are willing to move and there are opportunities elsewhere; in many countries the fate of those who have left the land to seek livelihoods in the cities is severe poverty. The second response cannot always rely on groundwater or streamflows being available. Where they are not, runoff agriculture may be the best option.

Much of the world's available freshwater runs away or evaporates before it can be used by agriculture or before it recharges groundwater; a large proportion of the waste could be prevented by SWC and runoff agriculture. Investment in these strategies has been comparatively neglected; yet such strategies should:

- cost less than large scale irrigation schemes;
- help reduce urban migration by improving or sustaining rural livelihoods;
- counter soil degradation;
- cut dependency because it can use local materials and reduces the need for food import;
- help recharge groundwater, reduce flood damage caused by uncontrolled runoff, improve regularity and quantity of streamflow;
- help those seeking to conserve flora and fauna.

RUNOFF AGRICULTURE: DEFINITION AND CHARACTER

Runoff may occur as overland flow, intermittent streamflow, or subsurface runoff (see the Glossary at the end of this book). I differ with attempts to restrict the definition to practices in semi-arid regions because runoff agriculture techniques are used or could be usefully adopted in other environments. Some argue that runoff agriculture should not involve storage, streams or springs; however, in practice most strategies use streams and springs when available, as well as collect other runoff. A reasonable working definition is that 'runoff agriculture provides moisture by collecting surface or subsurface runoff where other sources are likely to be too costly, unsustainable or damaging' (National Academy of Sciences, 1974). Stern (1979, p32) provided another usable definition of runoff agriculture (including cropping, herding and forestry) as 'the practice of concentrating surface runoff for cultivation'. Runoff agriculture is a generic term applied to cultivation, pastoral, forestry or conservation techniques that rely upon tillage or planting patterns, bunds, terraces, and other structures to delay and retain runoff, and to increase infiltration and counter soil erosion. Runoff agriculture draws upon the techniques of SWC and uses water harvesting, seasonal or ephemeral streamflows or floods, and snowmelt, rather than permanent riverflow or groundwater (Dupriez and De Leener, 1992). Runoff agriculture effectively uses moisture which would otherwise run to waste.

Rain-fed agriculture (cropping, herding or forestry) depends upon exploiting the precipitation received by the land for crops, fodder or trees. Runoff agriculture also makes use of water from adjacent land or more distant ephemeral sources. Runoff agriculture can harvest and concentrate ephemeral flows so that crops, fodder, trees or natural vegetation may be better grown

---

Box 1.2 ADVANTAGES OF RUNOFF AGRICULTURE OVER MAINSTREAM IRRIGATION

Runoff agriculture:

- can be adopted by the poor;
- has materials and methods that make it suitable for remote and difficult areas (marginal situations);
- is cheap to establish so is likely to mean much less dependency and release of funds for other development uses;
- makes use of local materials so is easy to adopt and maintain, and reduces dependency;
- implies conservation of soil and moisture;
- has less likelihood of causing waterlogging, salinization, pollution of rivers, groundwater or other waterbodies through return flows, and is unlikely to promote human diseases often associated with mainstream irrigation;
- has high potential for sustainable development;
- contains a diversity of approaches and techniques that help safeguard against large-scale failures;
- is cheap and can grow subsistence crops.

---

than environmental conditions would otherwise allow (Lovenstein et al, 1991). It can also feed water storage tanks or cisterns for livestock, domestic supply or irrigation. Runoff control can mean agriculture where rain-fed methods are insecure or impossible, better yields and more security of harvest or pasturage. Even one or a few collections of water a year can, through SWC, considerably improve biomass production or the establishment (or re-establishment) of vegetation cover. Runoff agriculture can be adopted by small scale landusers and by larger commercial producers. It may offer better opportunities for rural employment and more chance of sustainable development than large scale mainstream irrigation or mechanized rain-fed agriculture, and it should be easier to implement as extensive tracts of good quality land become scarce (Turner, 1994) (see Boxes 1.2 and 1.3).

Human and livestock populations have grown rapidly in a number of semi-arid and seasonally dry regions and look likely to increase. Today around 600 million depend upon rain-fed agriculture, using roughly one third of the world's total available land. Many of these people suffer from drought and land degradation, but aid is unlikely to come in the form of piped or canal supplies because the populace lives in remote or rugged terrain, or where soils are poor, and may be scattered too thinly to justify infrastructure costs even if river or groundwater supplies were available and could be channelled towards them.

In semi-arid and seasonally dry regions food and fuelwood production must be more secure and accessible to the poor. What is needed are ways to grow more food and fuelwood with less water as components of productive, sustainable, rural livelihoods (Arnon, 1981). What is required is 'thrifty irrigation': methods that are simple and can be quickly spread, sustained with local

> ## Box 1.3 Advantages of Runoff Agriculture over Rain-Fed Agriculture
>
> Runoff agriculture:
>
> - has the potential for reducing soil degradation and supporting better land husbandry;
> - is more likely to allow sustainable development;
> - may allow improved security of harvest;
> - may result in improved yields per crop harvest, perhaps also more crops per year;
> - has the potential for production in areas that are naturally unfavourable for rain-fed and perhaps also irrigated agriculture;
> - has the potential for crop diversification;
> - might be a way for rain-fed agriculture to respond if there is global climatic change towards less moisture availability;
> - can improve quality and quantity of streamflow, increase groundwater recharge, and reduce risk of landslides and floods.

materials and available labour and managed with easily mastered skills, and which are socially, economically and environmentally appropriate (Postel, 1992, p99). Unfortunately, many of the efforts to promote runoff agriculture and SWC measures have been clumsy and often top-down, have used tractors or components that must be shipped in, and have often alienated local people. For one or more of these reasons SWC and runoff agriculture have occasionally failed to operate well, sometimes even exacerbating rates of soil degradation and agricultural decline.

Runoff agriculture and SWC can be valuable in dry or humid environments and tropical to temperate regions (see Figure 1.1). However, in more humid conditions the value of SWC to control damaging overland flow is likely to be more important than improving moisture availability (Alconada et al, 1995). Runoff agriculture can offer a means of avoiding salinization (although there are situations where it can cause problems) which other approaches might provoke and has much to offer developed as well as developing countries. However, without incorporating water storage measures, runoff agriculture is likely to involve some degree of risk of crop loss.

In parts of India, the Far East and the Middle East innovative runoff farming techniques were developed over 5000 years ago. There are irrigated rice terraces at Bananue in the Philippines which are probably more than 3000 years old, which means that they are among the oldest sustained agriculture systems in the world. Recently, however, some have started to break down (see Chapter 5) (Stern, 1979, p15). In the past people have often relied heavily on runoff cultivation. In Peru before the Spanish Conquest there were over one million hectares; nowadays less than one fifth of that is in use but the potential for re-establishing it is there.

As new lands were settled by Europeans from the 16th century onwards, land degradation became enough of a problem for a number of colonial

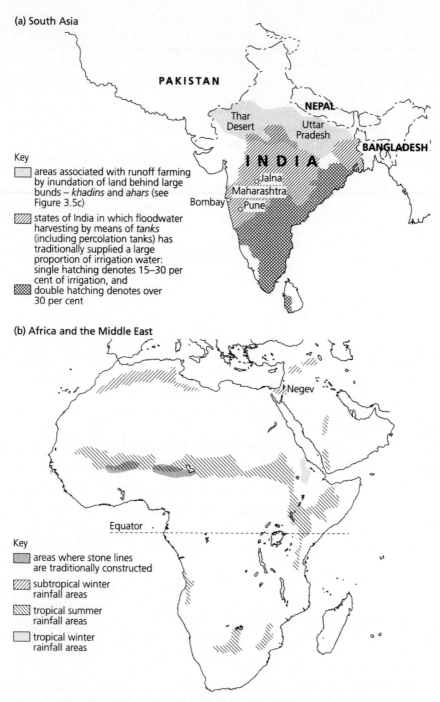

(a) South Asia

Key

☐ areas associated with runoff farming
by inundation of land behind large
bunds – *khadins* and *ahars* (see
Figure 3.5c)

▨ states of India in which floodwater
harvesting by means of *tanks*
(including percolation tanks) has
traditionally supplied a large
proportion of irrigation water:
single hatching denotes 15–30 per
cent of irrigation, and
▨ double hatching denotes over
30 per cent

(b) Africa and the Middle East

Key

▨ areas where stone lines
are traditionally constructed

▨ subtropical winter
rainfall areas

▨ tropical summer
rainfall areas

☐ tropical winter
rainfall areas

Source: (a) Part based on Pacey and Cullis (1986), p39, Fig 2.3; (b) Part based on Pacey and Cullis (1986) p128: Fig 6.1 and p162: Fig7.2

Figure 1.1 Areas of runoff agriculture or with potential for runoff agriculture in South Asia, Africa and the Middle East

authorities to introduce SWC measures and to pass land protection laws; indeed, by the late 19th century, such measures were widespread (Grove, 1995, p55). In the US several states (including Alabama, North and South Carolina, Florida, Mississippi and Virginia) promoted SWC well before the 20th century, although often in an insensitive, authoritarian and ineffective manner. The US midwest Dust Bowl tragedy of the 1930s (Worcester, 1979; Bonnifield, 1979) prompted the formation of the US Soil Conservation Service (part of the US Department of Agriculture) in 1935 by reorganizing the already established Soil Erosion Service (Helms and Flader, 1985). Unfortunately, SWC methods have not always been sensibly appraised before implementation, and past colonial extension activities may also have turned people against measures, even decades after independence.

2   Soil and Water Conservation (SWC)

This chapter examines soil and water conservation (SWC), focusing on its value for good land husbandry and for runoff agriculture. Coverage includes: indigenous SWC (some term this traditional SWC or view it as part of ethno-engineering); more recently developed techniques; and the value of SWC for crop production, pasture, forestry (especially in drylands and steep lands), land rehabilitation, and green development (Cheleq and Dupriez, 1988; Sandys-Winsch and Harris, 1994). Subsurface runoff and its conservation is dealt with in Chapter 3; the control and use of spate and flood flows is discussed in Chapter 4.

Agricultural yields can be raised through: crop improvement (better varieties); obtaining more than one crop a year; increasing moisture availability; raising soil fertility; and by reducing losses to pests, diseases, weeds and in storage or processing. Huge sums of money have been invested in improving crops, developing agrochemicals, mechanization and mainstream (often high tech) rain-fed agriculture and irrigation. These developments have been applied mainly in localities with good soil, adequate rainfall or irrigation supplies, good access to markets, and mainly to large-scale, commercially orientated agriculture. Rain-fed farming can be upgraded through irrigation or the application of chemical fertilizers but these require investment and are difficult to support in remote and rugged areas – it is in precisely these areas where SWC has potential. As the availability of water supplies becomes more difficult and costs increase, and as population in marginal steeplands and drylands grows (in Java alone there are over 91 million smallfarmers in upland areas), runoff agriculture will become an increasingly attractive alternative, or even the only practical approach to feeding and employing people and providing fuelwood (National Academy of Sciences, 1974).

There is a need to counter the worryingly high rates of soil degradation in developed and developing countries (Boardman et al, 1990; Pimentel, 1993). The conservation of soil and water are largely interdependent. The goals of SWC are to keep moisture and soil in situe and, if need be, to dispose of excess water. Water harvesting and spate irrigation discussed in later chapters also seek to intercept runoff and to collect, channel and use the water.

In addition to collecting and conserving available runoff where crops or other vegetation can benefit, or channelling supplies to storage tanks, SWC should address the problems of rain-splash erosion, sheet erosion, rill and gully erosion, and the risk of mass movement (landslides and mudslides). The primary tasks of SWC are to fight land degradation and to increase productivity of the land for crops, pasture or trees. SWC can maintain, or even improve, soil fertility by trapping and holding organic matter and fertile sediments, although moisture conservation and runoff control are usually more immediate goals (Srivastava, 1993; Critchley, 1991). It is probably more accurate to say that SWC seeks to conserve water (or soil moisture) and control water flows in order to protect the soil – so it should perhaps be abbreviated to WSC (Hudson, 1992, p118). It is also difficult to say exactly when SWC becomes water harvesting (see Chapter 3).

SWC can modify steep slopes to make their safe use possible (Jodha, 1990; Vincent, 1992; Yoder, 1994). A great many people seek livelihoods from steep land, and some countries have little that is level; for example, 75 per cent of Jamaica's cultivation is on slopes steeper than 18 per cent (Moldenhauer and Hudson, 1988, p33; Cracknell, 1983), as is much of the agriculture of Nepal, Bhutan, Afghanistan, Tibet, Burma, Pakistan, the Hindu Kush, Java, Taiwan, parts of the Middle East and North Africa, Andean South America, and many islands (Gumbs, 1993). People may have settled steep lands to avoid conflict or competition with others, to escape lowland diseases or climate, because they can develop a diversified livelihood strategy by exploiting different altitude zones, or for the reason that there is little other unsettled land available. Some of these steep lands have large and growing populations (the mountain areas of Rwanda may support over 800 persons per square kilometre – IFAD, 1992, p23) who need to adopt effective SWC and runoff agriculture. Some steep lands have carried much greater populations in the past but there has been a breakdown of SWC strategies, or people have simply moved. In the highlands SWC methods may have the added advantage of reducing or avoiding frost risk. SWC is especially valuable in highlands because land degradation there will often have serious (off-site) impacts on lowlands, such as landslides, silting-up of reservoirs and channels, erratic streamflow, and reduced groundwater recharge because highlands have suffered reduced infiltration.

SWC can be a way to give rural people better livelihoods, a way to rehabilitate degraded land (IFAD, 1992, p9, argues its value for this, especially for sub-Saharan Africa), and it might provide peri-urban areas (squatter settlements near cities) with a means for people to grow some food and commodities (see also Chapter 7). SWC can be useful for conserving flora and fauna. It is used in rain-fed agriculture as well as in runoff agriculture, and is employed in the control of flooding, watershed management, forestry and agroforestry (Troeh et al, 1980; Srivastava et al, 1993; Partap and Watson, 1994). IFAD (1992, p9) saw potential in SWC for fighting gender inequality since it can be an accessible means for women to get better supplies of food, fuel or income.

SWC strategies and techniques must be selected to suit the environmental, social and economic needs of a given situation (Dehn, 1995; Hein et al, 1997) (see Box 2.1). In addition to effectively and sustainably conserving soil and water, it is best to seek an approach which:

- involves minimal waste of land;
- requires reasonable inputs of labour and investment to install and maintain;
- minimizes hindrance to cultivation and possibly mechanization.

In some situations there is a tradition of SWC; elsewhere, indigenous SWC may be mixed with introduced methods of SWC (Van Dijk, 1995), and sometimes SWC has been adopted without having been used before. It must be stressed that, over the last 50 years or so, success with SWC has been patchy and some of its promoters tend to play down failures and stress potential (Critchley et al, 1994). SWC has great potential but it must be planned and implemented with great care and an awareness of its weaknesses. Published experimental data is derived from a diversity of research methodologies and different local conditions, so general conclusions are unwise but common. What works for one crop and style of landuse may not work for others.

There is a huge diversity of indigenous and more recently developed SWC approaches. Many different people have evolved techniques and strategies to suit very diverse local conditions (slope, soil, precipitation, aspect) over a long period (Dregne, 1986; Doolette and Smyle, 1990). Implementing the most appropriate SWC approach for the physical conditions is not enough; it is important that those promoting SWC recognize that the economic goals of poor farmers may be very different from those of commercial agriculture. Smallholders cannot wait long for returns and do not enjoy access to credit or benefit from economies of scale (Stocking, 1988; Magrath, 1990). Indigenous SWC can act as a starting point for developing strategies to sustain and improve agriculture and encourage good land husbandry (Reij, 1991; Kerr, 1992; Critchley et al, 1994; Hagmann and Murwira, 1996; Reij et al, 1996). Caution is needed because strategies which work in one locality may not successfully transfer to another, apparently similar, situation, although with modification and care they might. Selecting appropriate SWC approaches for a given situation requires adequate information on runoff soil characteristics and erosion risk (Lal and Russel, 1981). SWC should be carefully 'fitted' to local conditions and needs (see Figure 2.1).

Runoff may be generated by rainfall, or (sometimes in dry regions almost solely) by vegetation trapping fog and mist, or by spring and early summer snow melt. Runoff agriculture can utilize natural snow melt by diverting ephemeral or seasonal streams, but it is also possible to use SWC techniques to improve on nature, using fences, forest planting, vegetation barriers and crop patterns to trap snow which would otherwise blow away. When this melts it can moisten planting areas or be collected and channelled to where it is needed. Snow collection may be in highland areas, providing runoff for warmer lowlands, or may take place on crop land itself in cold-winter regions. It is possible to use SWC techniques in the upper sections of a watershed to trap snow or fog and mist, and to improve infiltration of precipitation so that

(a) based on field studies in Benin

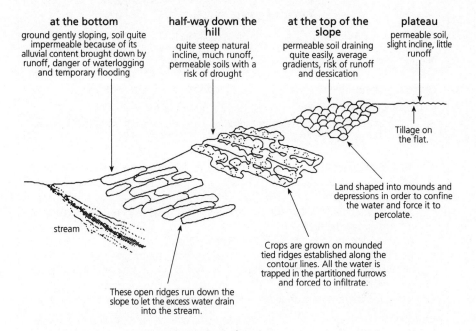

at the bottom
ground gently sloping, soil quite impermeable because of its alluvial content brought down by runoff, danger of waterlogging and temporary flooding

half-way down the hill
quite steep natural incline, much runoff, permeable soils with a risk of drought

at the top of the slope
permeable soil draining quite easily, average gradients, risk of runoff and dessication

plateau
permeable soil, slight incline, little runoff

Tillage on the flat.

stream

Land shaped into mounds and depressions in order to confine the water and force it to percolate.

Crops are grown on mounded tied ridges established along the contour lines. All the water is trapped in the partitioned furrows and forced to infiltrate.

These open ridges run down the slope to let the excess water drain into the stream.

(b) an example of comprehensive SWC in the Sahelian zone (West Africa)

Limit splash and create infiltration zones in all the cropping fields by means of suitable farming practices and perennial plantations.

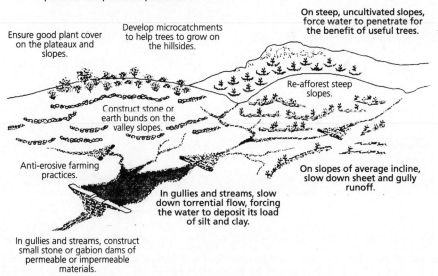

On steep, uncultivated slopes, force water to penetrate for the benefit of useful trees.

Ensure good plant cover on the plateaux and slopes.

Develop microcatchments to help trees to grow on the hillsides.

Re-afforest steep slopes.

Construct stone or earth bunds on the valley slopes.

Anti-erosive farming practices.

On slopes of average incline, slow down sheet and gully runoff.

In gullies and streams, slow down torrential flow, forcing the water to deposit its load of silt and clay.

In gullies and streams, construct small stone or gabion dams of permeable or impermeable materials.

Source: (a) based on Dupriez and De Leener (1992), p157: Fig 250; (b) based on Dupriez and De Leener (1992), p165: Fig 264

Figure 2.1 Fitting SWC techniques to the landscape (especially to soil and slope)

BOX 2.1 REQUIRED INFORMATION BEFORE PROCEEDING WITH SWC

Before SWC is undertaken, efforts should be made to establish the types of erosion and the causes (not symptoms), by asking the following questions:

Physical Questions
- Does splash erosion occur; where? Does a hard crust form on or in the soil?
- What pattern and quantity of precipitation can be expected; what is the probability of a given runoff quantity?
- What are the watershed characteristics?
- What type of erosion is observed on the plot – sheet; rill; or gully? The extent?
- What damage has already been caused by the runoff (gullies, landslides, water-logging)?
- What is the debris load of the runoff (is it sandy; silty; does it contain material rich in plant nutrients)?
- At what times of the year are the erosion and runoff most acute?
- What is the length of the growing season; what time of year does most crop growth take place (winter or summer growth); are most rains in winter or summer?
- Are some current or planned crops and farming practices more erosive than others?
- How are crop yields affected by observed phenomena?
- Are there threats: droughts; flash floods; landslides; frosts, pests; earth movements?
- What is the peak discharge (see Glossary)?
- Are there soil characteristics that need to be noted?

(a) sheet erosion (may be difficult to see in practice)   **(b) rill erosion**

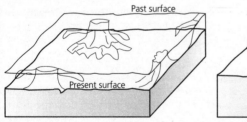

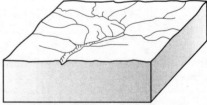

(c) gully erosion

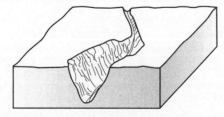

Socio-economic Questions

- Who owns the land? Who farms it? Is it in the interests of the owners to combat erosion and runoff?
- Who enjoys ownership of trees and hedges? Who is allowed to plant and exploit them? What are trees and hedges used for (eg fodder, woodfuel)?
- If land improvements (such as SWC) are carried out, who will benefit? Who will own them? Will land rents increase?
- Do local people have a tradition of SWC; can they cooperate and organize?
- Is conflict likely between locals and nearby groups if there is improvement? Is there any chance of regional unrest?
- What is access to market like (communications, distance, problems with transport, middle men)?
- Have there been past failed attempts at improvement?
- What do people need and what do they want?

Sheet erosion is often difficult to recognize in practice; gully erosion is obvious and may divert attention from (i) and (ii).

Source: various; some data from Dupriez and De Leener (1992, p156, Table 249); drawings (i), (ii), (iii) above by author

streamflow becomes less erratic and more substantial and groundwater is better recharged – thus runoff control can maintain streams and groundwater for a whole watershed (see Box 2.1).

It is possible to estimate the likely soil loss for a site under given landuse with empirical equations, such as the universal soil-loss equation (USLE) and its derivatives, through modelling or through field simulation experiments. Predicting runoff characteristics is a complex field (Finkel, 1986, pp9–11; Webster and Wilson, 1966, p125) that has generated a large literature. Once runoff is estimated it has to be compared with soil characteristics: erodibility, erosivity, infiltration capacity and likely agricultural or forestry needs. Runoff and soil loss estimation mainly relies on modelling (for runoff it is the FAO water balance model) or empirical formulae approaches (for soil loss under given environmental conditions and landuse, the USLE, the revised soil-loss equation (RUSLE) and other derivatives are used: see Glossary) (Wischmeir and Smith, 1960; 1978; Greenland and Lal, 1977; Morin et al, 1984; Renard et al, 1991; Victor et al, 1991; Mellerowicz et al, 1994; Pretty et al, 1995; Ferguson, 1996). Soil erosion is a function of many factors, including exposure, vegetation cover, surface litter, slope, rainfall intensity, soil characteristics, and whether there has been disturbance.

Wind erosion is mainly a problem of drier regions; in these areas SWC measures may focus on reducing dry season or winter wind erosion (when the land is left exposed between crops), as well as damage by flowing water, using wind breaks, cover crops, stubble mulch, etc) (Hagen, 1991). Water erosion is especially likely where soil becomes sealed or saturated and there is enough precipitation to cause surface runoff or subsurface flow at depths of less than a few metres (subsurface runoff).

SWC structures, such as terraces, stone lines, and vegetative barriers, counter runoff but they do not reduce rain-splash or soil compaction, so it is vital that those who coordinate soil and water management consider the soil quality and erosion protection of the planted plot and related structures (for example, it is important to consider land husbandry and not just the installation of SWC measures) (Gerasimenko, 1992). Ideally, SWC should not lead to soil degradation (such as soil compaction, acidification, waterlogging or salinization). For a review of such problems, see Unger and Cassel, 1991.

It is important with SWC not simply to concentrate on physical parameters, techniques and economics, which often seems to have been the case, but to consider all relevant dimensions of rural livelihoods, such as sustainability and organizational, social, political and community issues (Constantinesco, 1981; Halbach et al, 1988; Vincent, 1995, p13). SWC must address *causes* of soil degradation and moisture loss, not *symptoms*. Even if physical, technical and economic potential seem satisfactory, socio-economic factors could prove a problem; for example, a history of feuding between local people may prevent vital cooperation. There are situations where if care is not taken SWC will have a damaging effect on wildlife. However, occasionally the opposite may be the case and conservationists might profit from a study of its potential. Attention must not just focus on the areas where SWC measures are applied, but also on surrounding land and potential off-site impacts – for example, runoff from a path above well-made and adequately maintained SWC measures may cause damage. Studies in Ecuador by Harden (1992) suggest paths may generate more runoff erosion problems than the terraced fields and pastures they serve! Harden (1992) also warned that there might be a need for abandoned land and pastures to be considered along with cultivated lands, given the amount of runoff they can yield and the damage this might do if uncontrolled.

Because techniques which work in one area may not succeed when transferred to another, each initiative should be carefully planned. SWC must be pursued with great caution as it is a 'minefield for the unwary' (Stocking, 1988, p385). Too often efforts have been inappropriate or clumsy and the results disappointing and unsustainable, even damaging to people and the environment (see Chapter 6). In any given situation there may be several possible techniques and perhaps more than one means of constructing and managing each. Those promoting SWC and runoff agriculture must allow farmers to experiment and chose what works for them, not impose things (Kiome and Stocking, 1995). Better ways are needed to select the right SWC approach for a given situation. Gerasimenko (1992) has examined this for the European CIS; another possibility is to apply *expert systems* (see Glossary) to selection (see Box 2.2).

SWC efforts over the last 30 years or so have had some good and some disappointing, even seriously damaging, results; for a recent review, see Reij et al, 1996, who present case studies from sub-Saharan Africa. Failures have often, at least in part, been due to an insensitive approach, usually married to a failure to collect and heed local data. SWC has sometimes been seen as a way to boost 'backward' agriculture and raise yields, the data used to plan it has come from unrepresentative surveys and experimental plots and the planners

---

Box 2.2 The application of SWC

SWC may be pursued via one or a combination of:

- area closure – slow, but often effective, although it may well mean those excluded have to find alternative livelihoods (fencing can be expensive and prone to damage, and may hinder wildlife movement);
- conservation tillage;
- engineering or structural means such as terracing;
- vegetative measures such as grass strips.

The land on which SWC is used includes:

- non-agricultural land;
- marginal land;
- prime agricultural land;
- conservation areas;
- degraded land.

SWC may be used on land exploited in many different ways, including for:

- annual crops;
- perennial crops;
- forage;
- forestry;
- nature conservation.

---

neglect to consider other issues like sustainability. There have been recent improvements, in particular a growing shift since the 1970s from top-down application of SWC to more participatory bottom-up land husbandry which can draw upon SWC. Shortage of funds for investment, often as a consequence of structural adjustment programmes in developing countries and recession in developed countries, has helped stimulate interest in lower cost SWC and runoff agriculture, rather than in mainstream irrigation.

SWC is a very broad field which is not easy to adequately subdivide or classify. Hudson (1987, p49) and IFAD (1992, p39) tried, suggesting a subdivision into three broad categories based on precipitation conditions:

(1) *High rainfall SWC* (mean annual precipitation >1000 millimetres per year). The objective is to dispose of large quantities of runoff, prevent waterlogging and conserve soil (see also Sheng, 1989). Where there is heavy seasonal precipitation the excess runoff may be collected from SWC structures and stored in tanks or cisterns for livestock, domestic or irrigation use during dry season or drought periods.
(2) *Medium rainfall SWC* (mean annual precipitation 700 to 1,000 millimetres per year). The objective is to hold moisture in situe and conserve soil; excess runoff may be intercepted and stored in tanks or cisterns for

boosting yields or as security against crop loss. Reij (1991, p6) suggested the shift from harvesting to holding moisture in situe occurs where precipitation is greater than about 500 millimetres per year.

(3) *Low precipitation SWC* (mean annual precipitation 300 to 700 millimetres per year). The objective is to capture and concentrate runoff (runoff harvesting practices are usually undertaken where rainfall is between 100 and 700 millimetres per year), and to transfer it to a site where soil can be conserved (FAO, 1988). It may also help to grow drought-tolerant crops, trees or forage and to adopt practices such as mulching. In practice many regions have such variable rainfall from year to year that they may not fit into one of the previously listed categories. In such situations it may be possible to modify SWC methods from season to season – for example, opening the end of contour bunds before the wet season so that excess runoff can be shed to a waterway, reducing the risk of damage to structures or planting areas.

There are other possible subdivisions; Lal (1991a) divided technical options for erosion management into preventative measures (mulching; cover crops; conservation tillage; strip cropping; vegetative barriers, agroforestry and alley cropping; contour farming; groundcover management) and control measures (graded channel terraces; ridge tillage; contour bunds; waterways; drop structures and adequate aprons of pebbles or concrete to prevent damaging scour; diversion channels, gabbions; check dams, etc). Hudson (1987, p49) suggested SWC should be divided into, internal – where runoff is stored and used more or less at the site where the structures are built and crops planted; and external – runoff is diverted to some, perhaps quite distant, point (demanding careful control of potentially damaging channelled flows).

Indigenous SWC evolves to suit specific circumstances (see Box 2.2), and commonly a mix of methods are selected to suit changing slope gradients and local soils (see Figure 2.1). Techniques must often be flexible. For example, it may be necessary to selectively trap soil or debris in runoff to maintain or enhance fertility, rather than to stop all movement of debris; control of runoff designed to collect gentle flows may need to withstand occasional intense floods; and moisture collection and conservation must be balanced against risks of waterlogging.

It is important to know the pattern of precipitation as well as the amount before making any attempt to promote SWC or runoff agriculture. Efforts must be made to establish whether precipitation occurs as gentle showers or occasional intense storms, whether it falls year round, in summer, or during a cool season. The same mean annual rainfall well distributed through a time of year when air temperatures support plant growth without excessive evapotranspiration is much better than precipitation during a hot, dry summer or winter when it is so cold that nothing grows. Occasional, sudden, intense storms demand different treatment than frequent, more gentle precipitation; however, the annual precipitation values might be identical.

The value of SWC for improving land husbandry and supporting agriculture in difficult and marginal situations has increasingly attracted interest since the 1980s (see Moldenhauer and Hudson, 1988; Shaxson et al, 1989; Moldenhauer et al, 1991; Hudson, 1992). In many parts of the world growing human and livestock numbers, as well as other development pressures, are stressing established landuse, causing a breakdown of livelihoods and environmental degradation. With limited access to credit or other supports, and often living in marginal situations, rural folk might seek to survive by:

- using more land (but extending agriculture is increasingly difficult since suitable land becomes scarce);
- migrating to urban areas or overseas or to wherever employment opportunities exist;
- intensifying landuse, which is often the ideal solution and the most difficult for local people to achieve unaided.

The problem is how to intensify in order to improve livelihoods, sustain production and control environmental damage, and to do so with little help from outside the locality (Boserüp, 1965; Ghimire, 1993; Tiffen et al, 1994). SWC offers an accessible route to intensifying cropping, livestock rearing or forestry even for smallholders in difficult environments (Tato and Hurni, 1992). It is also possible to achieve SWC through agroforestry, which establishes a vegetation cover that does not return too much moisture to the atmosphere by evapotranspiration and which retards runoff, conserving soils and improving infiltration (Young, 1989).

SOIL AND WATER CONSERVATION IN PRACTICE

Well-constructed and managed SWC systems can function for centuries and are one of the few *proven* routes to sustainable agriculture. However, SWC benefits may be slow to materialize, dispersed and indirect. They may, at least in part, be enjoyed by people who do not labour or invest in the measures (for example, those off-site who are spared the siltation of water caused by runoff). Tracing beneficiaries and extracting payment for SWC could be difficult, so it is better to seek strategies that obviously and quickly reward those who labour or pay for them and who maintain them in the long term (Moldenhauer and Hudson, 1988, p26). While most countries pay lip service to SWC, it is seldom a vote-winning issue (Hudson, 1992, p32). If small-scale agriculturalists are to support SWC it must be affordable and they need to *see* significant and reasonably immediate benefits; soil conservation and sustainable development are important goals, but probably do not appeal to poor farmers (although land degradation is ultimately likely to cause marked declines in production and livelihood). Authorities often stress those goals, though it is better to emphasise and seek realistic, farmer-attractive agricultural improvement benefits (increased agricultural production; improved harvest security; reduced input of labour against returns; opportunities to diversify) (Critchley et al, 1992). Installation cost and ease of maintenance are important for large-scale commercial agriculture (Stonehouse, 1995), and it is crucial that SWC must be cheap and simple enough to be adopted by poor people in developing countries.

In some countries SWC may be paid for by the government (the cost recovered from taxes of various types), or aid agencies – but at present, more often than not it is the landuser who pays (Seitz, 1984; Hudson et al, 1993). Therefore, to make SWC accessible it should be based on local materials, practices and available labour (Barbier and Bishop, 1995). Strategies generally adapt better to challenges such as floods, earthquakes or landslides if they are based on simple, locally repairable structures which can be rebuilt easily, rather than seeking to survive extreme conditions; however, the latter is what developers often promote at considerable construction cost (Cosgrove and Petts, 1990). Where SWC is supported by grants or aid (such as food for work), farmers may be reluctant or unable to repair and maintain structures (aid incentives and subsidies are discussed in Chapter 6).

By improving soil moisture, SWC can make tillage and planting easier and less prone to delay (without SWC some soils can only be cultivated with the tools available to smallholders when rainfall has rendered the land workable – neither too wet, nor too dry); however, it may slow some agricultural tasks, for example by requiring ploughmen and draft animals to trek to a path and struggle down to the next terrace. Judging the success of SWC depends on small-scale agriculturalists' and observers' priorities – for example, erosion may be controlled and landuse sustained, but the terraces or whatever is used may reduce yields slightly. This may be attractive to an observer but probably not to the farmers. Improved yield, security and sustainability may not all be possible. Leach and Mearns (1996, p179) discussed attempts to compare terraced and unterraced land in Kenya (Machakos District), looking at yields, diversification of crops, soil conservation costs and benefits, and the diverse ways farmers paid for SWC.

A farmer is unlikely to operate in isolation; often slopes are worked by numerous smallholders. These smallholders need to get access to their plots and have to carefully dispose of excess runoff, or someone downslope or downstream will suffer. Coordination is likely to be necessary, and if there is no tradition, must be developed (see Chapter 6).

SOIL AND WATER CONSERVATION TECHNIQUES

Prevention of soil degradation (such as rain-splash and compaction) occurs mainly through maintaining protective vegetation cover (or mulching). *Control* of soil movement is by mechanical runoff control, typically earthen bunds, stone structures or vegetative barriers. Interest in vegetative barrier approaches has been increasing, but the approaches are mainly applicable in more humid environments; in drier situations, where plants grow poorly, mechanical SWC techniques predominate. Whatever methods are used, SWC is helped by maintaining soil fertility – part of good land husbandry.

There have been various attempts to assess the relative merits of SWC techniques – Sutherland and Bryan (1990) used catchment experiments in Kenya to compare different strategies. Most of these efforts have given approximate values that are difficult to compare with those collected in other localities. One such study is that of Dano and Siapno (1992), who examined the effectiveness of various SWC measures in reducing soil erosion and runoff, together with their acceptability to small farmers in steep highland areas of the Philippines. The results may not be universal but at least give some insight; for 30 per cent to 60 per cent slopes, bench terraces gave rapid results while contour hedges had a more gradual effect, but were cheaper and more attractive to farmers. The bench terraces reduced soil losses by about 80 per cent; stone lines were 78 per cent effective and contour hedges were 68 per cent effective (similar assessments have been made by Gicheru, 1994).

Soil and water conservation by agronomic techniques

Some mechanical agronomic SWC techniques are much used in landscape engineering, for instance to protect road embankments against erosion, and some are used by agriculturalists who might argue they are practising rain- fed agriculture. A number of the techniques prevent or reduce rain-splash damage, which is important for runoff and rain-fed agriculture. *Conservation tillage* is a generic term for the use of tillage techniques or mulching and groundcover for SWC. The field has generated a large literature (see Mannering and Fenster, 1983; *Soil & Tillage Research*, vol 27, nos 1–4 review tillage methods and SWC worldwide). In practice, more than one of the SWC techniques discussed in the following pages may be used in combination, which gives very different results (for instance, a tillage technique plus mulching and bench terracing). A given technique is likely to suit some conditions better than others but there can be considerable local variation of soil, so it needs to be adaptable, even over a small area (Hien et al, 1997).

Mulching can help slow runoff and improve infiltration, suppress weeds and deter pests, protect against rain-splash erosion, excessive UV radiation and heat damage to the topsoil, and help conserve soil moisture. Organic material mulches also break down to form useful compost, helping to sustain soil fertility. Mulching is a form of soil amendment (not to be confused with chemical treatments intended to counter saline or alkaline contamination – for example, the use of gypsum on salinized soil), which may be used on its own or combined with various other SWC techniques. On its own it is most effective on gentle slopes (Srivastava et al, 1993, p21).

Mulch may be made from many organic or inorganic materials (hedge or tree prunings; crop residue; straw; cocoa husks; peat; sand; gravel; dust; vermiculite; shredded tree bark or waste paper). Suitable materials are those which are cheap, locally available, easy to transport and apply, stay in place, do not prevent air reaching the soil or damage useful soil organisms, act to reduce evapotranspiration and improve infiltration, discourage weeds and pests, and break down slowly, adding to humus and soil nutrients – although care must be taken that decomposition does not lead to soil nitrogen loss (Barrow, 1987, p155). Mulch with a high albedo can help conserve moisture and protect the soil from heat damage by reflecting more solar radiation than dark soil.

Many countries have traditional mulching methods which use crop residue or straw (see Richards, 1985, p60 for details of some of those in sub-Saharan Africa). In Australia and North America tillage plus crop residue mulch (stubble mulch) is a common strategy, easily carried out by agricultural machines, which makes it attractive to large-scale cultivators (Freebairn et al, 1993; Allen and Fenster, 1986). Mulching may take the form of bands, rather than a continuous spread to protect vulnerable sites, or make the best of scarce mulch material – known as strip mulching (Sur et al, 1992). Mulching can be effective; for example, in Kenya, tree prunings' mulch was found by Omoro and Nair (1993) to cut soil loss by as much as 89 per cent and runoff loss by up to 58 per cent (compared with control plots). However, there may be situations where use of crop residue for mulch reduces the availability of fodder.

Mulching may rely upon non-organic materials: a 5- to 15-centimetre thick layer of sand, pebbles, gravel or dust spread over the soil is a common strategy. Although widespread, the real value of non-organic mulching has been debated; undoubtedly it protects the soil from UV damage, overheating and rain-splash erosion, and it may discourage weeds and some pests, but whether it reduces moisture loss (Wrigley, 1981, p88 was sceptical) or can trap dew is less certain. It is widely held that mulching improves infiltration by retarding surface flow rate (Jackson, 1977, p79).

Gravel and pebble mulching was used by several North and Central American Indian peoples before AD 1500 (for example, the Anasazi and Pueblo peoples of New Mexico); it also has a long tradition in North Africa and a number of other countries (Barrow, 1987, p157; Lightfoot, 1993; 1994). In Lanzarote (Canary Islands) farmers discovered volcanic ash made an effective mulch for microcatchments that grew vines, following an eruption in the 18th-century (Hall et al, 1979, p205).

Plastic sheeting is used for mulching in some areas, mainly by commercial farmers. Too expensive for smallholders, it also presents disposal problems once damaged; if scattered by the wind this can be a nuisance, although some plastics are now biodegradable. Recent developments adopted mainly by civil engineers and landscaping contractors are *hydro-seeding* and *rolled erosion control systems*. Hydro-seeding is the spraying of a mulch, or sewage sludge, or wastewater biosolids, or fibre-mat forming slurry enriched with seeds. Once applied, the excess water evaporates leaving a moist, soil-anchoring seedbed. Rolled erosion control systems are fibre-mats or sheeting – mainly used for construction projects, landscaping and land rehabilitation, and today too expensive to be widely adopted by smallholders. There are many types and various makes of these 'geotextile' mats and fibre-blankets – the main value for small-scale agriculture is in their potential for stabilizing sand dunes or slopes that threaten farmlands or villages or to reinforce steeper cut-off drains, waterways and earthen dams. Some are formed with natural fibres, such as jute, coir, kenaf, bargasse (sugar-cane waste); these biodegrade rapidly and weeds or grass take over erosion control. Others use synthetic fibres such as PVC, polypropylene or glass-fibre and may remain intact for decades (for a review see Sutherland, 1998).

Dissolving small amounts of certain polymers in irrigation water or spraying an emulsion on the ground may improve infiltration and reduce moisture loss. These *hydrophilic additives* can also be tilled into the soil as pellets. Starch copolymers such as polyacrylamide (under various trade names) have been quite widely used by landscaping companies and commercial foresters establishing trees or shrub shelter belts in poor rainfall or sandy-soil environments (Anon, 1982; *Unasylva*, vol 152, p38, 1986). However, this too is likely to be too expensive to be useful for smallholders but, again, it can offer authorities a quick way of stabilizing dunes or slopes that threaten farmers.

With free-draining soils it may be possible to retain moisture from runoff irrigation or rainfall within reach of crop roots by burying a plastic sheet when constructing the planting plot. Alternatively, the sheet can be introduced as a strip using a special plough. These measures can be effective ways of conserving moisture and reducing the loss of any fertilizers applied to the land. However, they are not yet cheap enough for general use; but may have potential for urban and peri-urban agriculture, especially if cheaper recycled plastic sheet becomes available (Gischler, 1979, p112). Many of these high tech methods degenerate quickly as a result of sunlight, animal damage and weathering, yet decay slowly enough to present a waste disposal problem. They are useful where cultivators can grow high-return crops for the market, and for establishing shelter belts or stabilizing moving dunes.

Thin *sprayed-on surface films* may be formed from wax, plastic, latex, oil emulsions, or bitumen, and have been used for land reclamation, stabilization following construction work, and to improve rainfall harvesting catchments. Their value for smallholders has been limited mainly due to high cost, limited life and, with some, the tendency to contaminate runoff water.

Soil amendment is a term also used for treatments that are intended to rehabilitate contaminated, especially salinized, soil – for example, spreading compounds such as gypsum on the ground surface. Where soil has been contaminated with hydrocarbons or other pollutants, bacterial cultures may be dug in to prompt *bioremediation*. If these treatments can improve infiltration, they have potential for SWC. *Green manuring* (the growing of crops which can be ploughed in to improve soil fertility and structure) chemical soil conditioners can help soil to catch and hold moisture and nutrients, and is much cheaper and more accessible to smallfarmers. Green manuring deserves much more funding to improve methods and disseminate its practice.

Fallowing is used by rain-fed and irrigated agriculturalists to allow soil fertility to recover, improve the moisture content, or to allow salts accumulated during cultivation to leach away after rainfall. There are traditional and modern forms of fallowing suitable for various environments, some of which may suit runoff agriculture. Deliberate planting of fallow crops to provide groundcover should discourage weeds and could possibly improve soil nitrogen. Such planted fallows often use legume species because these fix nitrogen effectively – for example, sesbania (*Sesbania sesban*); mucuna (*Mucuna pruriens*); cowpea (*Vigna unguiculata*). There are some, such as Payne et al (1990), Tanaka and Anderson (1997), and Lopez et al (1996), who doubt the value of fallowing for improving soil moisture, at least on deep, sandy soils and fine loams.

Where land slopes gently (up to 12 per cent or 7 degrees), adequate SWC may be achieved by conservation tillage (*contour tillage* or *contour farming*). At its simplest this consists of ploughing and planting crop rows along the contour (Lafond et al, 1994; Lal, 1994; Wiese et al, 1994). Planting along the contour, rather than up and downslope, should better conserve moisture and soil. Conservation tillage developed rapidly in the US following the 1930s Dust Bowl disaster. Some of its approaches have been developed in temperate environments and need to be adapted to fit other climates (Carter, 1994).

Conservation tillage (contour ridges, bunds or furrows) are useful techniques where the soil is crusted over or is lateritic, and the alternatives for improving infiltration generally involve considerable labour to break up the ground. For sandy clay–loam soils in southern Spain, Moreno et al. (1997) found conservation tillage highly effective in enhancing soil moisture recharge and soil water conservation, particularly in years of lower than average precipitation. *Contour strip cropping* is a widely used conservation tillage technique that involves alternating strips of crops with strips of grass, cover crops, and other plants. The approach can be used on slopes of up to seven degrees – the strip width being adjusted to suit the gradient. The strips can be used to establish a crop rotation. For added security and better infiltration, contour furrows or ridges are often 'tied' by hand or with a modified plough (Finkel, 1987, p46) (see Figure 2.2). *Ties* are earth barriers at intervals along a furrow, like rungs on a ladder, their crest lower than the ridges either side. They can, under the right conditions, help hold slight and moderate runoff and control heavier flows. *Tied contour tillage* (basin tillage) can be carried out with machinery and is often adopted by larger-scale producers (Morin et al, 1984). On slightly steeper ground or where there is more aggressive runoff *contour bunds* (slightly more substantial and higher than ridges) may be effective (if need be with ties).

Tied ridging (furrow diking or basin listing) consists of ridges, sloping gently across slope or sometimes downslope, with ties at intervals in the furrows between them. Crops are usually planted on the ridges. The approach is useful where there is periodic, heavy rainfall, the ties serving to slow excess runoff, and to trap water and soil; they may also counter erosion by preventing catastrophic overtopping of ridges which can result in gully formation. The technique is suitable for gentle slopes and heavy soils, and is usually adopted by farmers with access to tractor or ox-drawn ploughs which can be adapted to form the ties during ploughing. It is very labour demanding to do by hand and so unpopular with smallholders. Generally, tied ridging must be renewed annually or even after each harvest or heavy storms, although there is a form of 'no-till' tied ridging which typically needs renovation only every five years or so. Tied ridging has given mixed results: for example, results are good in the US and parts of India and Kenya, but there are cases where it has resulted in waterlogging and depressed yields and where ties have been overtopped in storms, leading to progressive failure and gully erosion.

Contour tillage methods are generally thought to waste less land than terracing and demand less labour input than terracing, and so can be attractive to small farmers.

(a) strip tillage, strip planting and strip cultivation

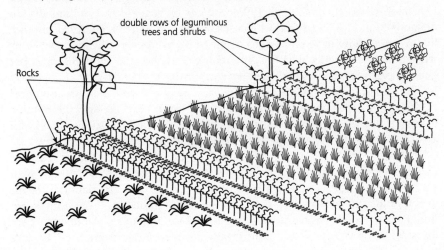

(b) ridge and furrows:

(i) contour ridge and furrows with ties

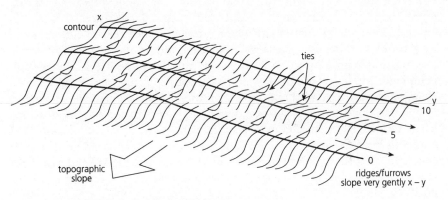

(ii) on gentler slopes ridges and furrows up–down slope with ties

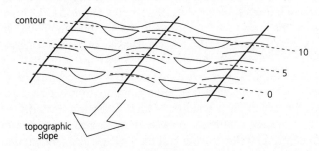

Figure 2.2 SWC techniques

SWC, if the right method is selected, can be a boon for agriculture, including smallfarmers (Biamah et al, 1993). Studies by Vogel (1992) in Zimbabwian drylands found SWC reduced soil losses, recording that conventional (mouldboard) ploughing gave soil losses of up to 9.5 tonnes per hectare per year, while conservation tillage with ox-ploughing could be as low as 0.5 tonnes per hectare per year (traditional hand-hoeing in the region gave quite high soil losses, something confirmed by various studies elsewhere). Benites and Ofori (1993) found that conservation tillage may have very site-specific results, making it unwise to overgeneralize. It is better to make pilot studies in a given area before undertaking full-scale developments. A factor which must be considered is the lifetime of structures – studies suggest that, at least on clay–loam soils, the effectiveness of conservation tillage decreases rapidly with successive rainstorms unless it is reformed (Lal, 1991b; Mandal et al, 1994; Shafiq et al, 1994). However, given the worrying losses of soil and moisture under many existing agriculture systems, and the impacts of eroded material on streams and reservoirs, almost any improvement from SWC is very welcome; the problem is providing the labour needed (Tagwira, 1992).

Tillage may cause soil changes which result in land degradation. Compaction or crusting can occur, sealing the soil surface. It can also contribute to the formation of an impermeable pan just below plough depth (a plough sole – a problem with some oxisols); the result is reduced infiltration leading to waterlogged topsoil, reduced soil moisture storage and groundwater recharge, and runoff which can cause sheet, rill and gully erosion. *Deep tillage* may break up subsurface pans and reduce the problems of waterlogging and damaging runoff. However, this may require heavy plough equipment and may sometimes cause unwanted problems, one of which, some claim, is excessive loss of soil moisture. In southern Zimbabwe, Hagmann (1996) describes the use of an ox-drawn plough to form contour bunds which have exacerbated (as local farmers feared), rather than cured, rill erosion. Had the locals been heeded, land degradation might have been avoided. Hagmann concluded that the transfer of know-how by Westerners can fail and that there is a need for flexible approaches which are established gradually through trials involving the farmers.

Tillage experiments show some variability of results. To generalize, SWC is beneficial, but any approach must be assessed before being applied to a given situation (for a review of tillage and SWC methods in the Caribbean, Australia, South Asia, and South Africa, see the 1993 special issue of *Soil & Tillage Research*, vol 27, nos 1–4). There may be times when SWC methods fail to adequately protect – for example after harvest or grazing if little crop residue is left; during and after tillage until crops or other vegetation grows enough to give protection; if people collect crop residue as fuel; following drought or bushfires and in dry environments where it may be necessary to avoid dense planting. The problem of vulnerability during and after tillage has generated a range of techniques (some terms overlap): strip tillage; strip cultivation; stubble tillage; stubble mulch; trash farming; minimal tillage; reduced tillage or low tillage; direct drilling; and zero tillage (no till). All of these practices seek to reduce or avoid exposure of the soil to erosion by runoff or wind while not hindering infiltration. They also return crop residue to the soil, helping to maintain soil nitrogen and thus fertility. Zero tillage and retention of crop residues (such as stubble) can lead to more insect problems and transmission of fungal and other crop diseases. Weeds that are likely to compete with crops or transpire soil moisture can be removed by using a scythe, power-strimmer, herbicide, fire or hand weeding.

Minimum tillage is generally understood to be one preseeding or preplant-ing operation; conservation tillage may well involve two or more tillings, before and after crop growth. Direct drilling, zero tillage and minimum tillage involve seeding without significant soil disturbance between crops (which may lead to oxidation of soil carbon and loss of fertility). This may be achieved by just cultivating a narrow strip or even inserting seeds with a special plough-blade (which usually demands especially powerful tractors). Stubble tillage, trash farming and stubble mulch aim to keep as much crop residue as possible covering the soil between harvest and the development of adequate cover by the next crop (Lal, 1983). These methods may be especially useful where rainfall is intense and soil is vulnerable, and where cold and dry winters or dry summers follow tillage and planting. Because manual labour and some draft-animal tillage cannot cope well with all of these methods, they have mainly been adopted by larger-scale cultivators with access to mechanization – for example, in parts of Brazil and Australia. However, there are some indigenous forms of reduced tillage.

One that is widespread is the sowing of beans into cereal stubble. Some controversy over the benefits of minimum or zero tillage has been generated since the 1940s. Tanaka and Anderson (1997) made an assessment of the relative value of no till, minimum till, stubble mulch and winter fallowing for improving the moisture storage of a fine loam in the Central Great Plains, US – they found no till and minimum tillage most effective in improving moisture storage (by up to 12 per cent). Radford et al (1995) made similar comparisons for the semiarid subtropics and found conservation tillage greatly increased grain yields by improving soil moisture storage.

For some soils the opposite approach to zero tillage might work, especially where the land is impermeable or pans form and hinder infiltration. On deep sandy loams (luvisols), deep tillage may be used to increase infiltration and drainage to encourage root penetration, thereby boosting crops and reducing damaging runoff flows. Deep tillage of such soils is likely to require mecha-nization or heavy draft animals, and so is not really suitable for smallholders.

Broad bed and furrows (raised beds) (also discussed as a wetland strategy in Chapter 4) are an ancient technique, especially useful for regions with deep vertisols (sticky, clay soils that harden when dry) subject to heavy rain. These soils are often difficult to cultivate unless the time is right, which is just after adequate, but not excessive, rain. Consequently, rain-fed agriculture may face delayed planting which reduces the cropped area and may by shortening the growing season increase chances of a failed harvest. As a result the rain-fed farmer faces insecurity, and, because runoff may not infiltrate enough, the land may shed runoff before groundwaters recharge, eroding the land and causing streams to flood.

Broad beds and furrows can make cultivation easier, reducing the need for careful timing of tillage and aiding drainage via the furrows. The land can be made more productive and crop diversification may be possible. By reducing erosion agriculture becomes more sustainable and gives more secure harvests than simple rain-fed agriculture. With grassed furrows erosion is better controlled (Astatke et al, 1989), and if these feed into storage tanks, supplementary irrigation of the catchment (by motor pump or ox-drawn water cart and sprinkler) or of land near the tank is possible (see Figure 2.3). The International Centre for Research in the Semi Arid Tropics (ICRISAT) has developed such a strategy for gentle slopes (0.4 per cent to 0.8 per cent) (Manassah and Briskey, 1981, p361).

Broad beds, raised beds and terraces can modify microclimate enough to help agriculture, especially where there is marked radiation loss at night, leading to frosts (as is the case in some tropical uplands). In Ethiopia's highlands, the International Livestock Research Institute (ILRI) has effectively promoted broad bed and furrows, made with an animal-drawn shaping device, to improve crop and forage production and combat soil degradation on vulnerable vertisols (El Wakeel and Astarke, 1995).

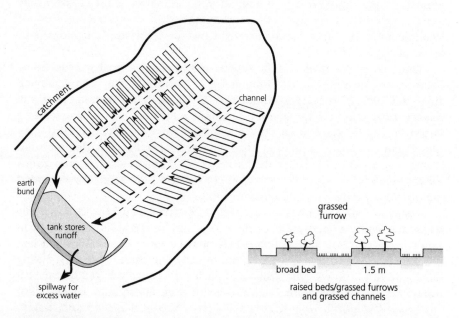

Note: schematic, not to scale.
Broad beds are formed by cut-and-fill as furrows are dug. Water from the storage tank can be used for irrigated crop downslope or can be applied to broad beds to augment rainfall. For the latter, a bullock-cart with sprinkler-bar can be used with the wheels running in the drains and furrows to minimize soil compaction. Method developed by ICRISAT (Hyderabad, India).

Source: original development by ICRISAT, India

Figure 2.3 Broad beds, grassed furrows and grassed channels draining to grassed drains that feed runoff to a storage tank

Soil and water conservation by mechanical techniques

There is not a very clear distinction between agronomic and mechanical techniques; however, Hudson (1992, p153) suggested it might make sense to group all mechanical SWC measures (for instance, ditches, bunds and terraces) which act as cross-slope barriers. Alternatively, mechanical SWC could be subdivided into alternative or appropriate methods which use local materials and skills, and engineering methods (high tech), that require an input of skills and material which are often not locally available (Schwab and Frevert, 1981).

Trash lines and *stone lines* offer advantages over earth banks or terraces which demand a lot of labour to construct and may be damaged by people's trampling and livestock poaching and by severe runoff, especially if not built to a high standard. For any sealed structure there is a danger of progressive failures: one structure after another failing until a serious gully appears. The risk of overtopping and progressive failure may be lessened if structures are semipermeable, for example: lines of trash (brushwood, crop residue or straw) or of uncemented, but firmly bedded, stones laid along the contour. All of these structures can trap moisture and debris and safely leak or spill excess water without needing expensive, exotic materials or carefully engineered spillways. They are thus ideal for smallfarmers, even in remote areas, because they use local materials, are low cost and are often easily adopted.

Stone lines (*diguettes*) are effective at slowing runoff where soils have poor infiltration qualities (see Figure 2.4a). They are also an excellent way of disposing of unwanted stones found as the farmer cultivates – turning an impediment into a resource. They have proved very effective for SWC in parts of Africa and have been found to significantly increase cereal yields (Hulugalle et al, 1990). They have been linked with Oxfam; however, although promoted and spread since the 1970s by Oxfam and other aid agencies and NGOs, they were not 'invented' by Oxfam. There is a long tradition of stone lines in West Africa, especially, in Niger and Burkina Faso (IFAD, 1992, p26; Atampugre, 1993). As with other contour ditches, terraces and similar approaches, stone lines must be accurately aligned along contour.

The best-known stone line success story is that of Yatenga Province, Burkina Faso, where, in the 1970s, Oxfam helped spread 'forgotten' stone lines and *zai* – traditional planting pits (see later in this chapter), to over 400 villages. An important part in this extension work was played by introducing the water-tube level, used to aid farmers to place stone lines exactly along the contour (water-tube levels and other appropriate levelling devices are discussed in Chapter 6). Stone lines have been found to increase crop yields between 30 per cent and 60 per cent within one year (and as soil and moisture accumulate should give even better long-term results) and to offer better security against crop failure in dry years than preexisting rain-fed agriculture. By the late 1980s over 8000 hectares had been treated in Yatenga Province and villagers were spreading the approach themselves (Postel, 1992, p116). For a comparison of the qualities of stone lines and earth bunds see Table 2.1.

(a) stone lines along contour (i = plan; ii = x-section)

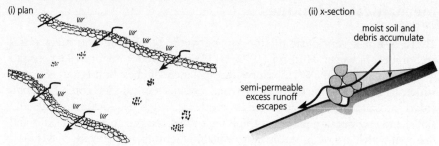

(i) plan

(ii) x-section

moist soil and debris accumulate

semi-permeable excess runoff escapes

(b) contour stakes (*clayonnage*) (i–v x-section as time progresses)

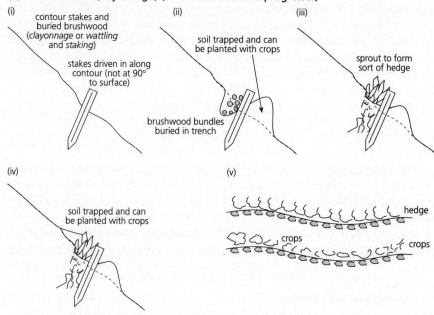

(i) contour stakes and buried brushwood (*clayonnage* or *wattling* and *staking*)

stakes driven in along contour (not at 90° to surface)

(ii) soil trapped and can be planted with crops

brushwood bundles buried in trench

(iii) sprout to form sort of hedge

(iv) soil trapped and can be planted with crops

(v) hedge

crops

crops

(c) contour bunds (i = ideal – better to place bund upslope to reduce erosion and catch soil in ditch; ii = unsatisfactory – soil washed off the bund is lost downslope, and poses a greater risk of gullying)

(i)

(ii)

Figure 2.4 SWC techniques

Table 2.1 Comparison of stone lines and earth bunds

Stone Lines	Earth Bunds
permeable	+/– impermeable
stone lines on gentle gradients or walls	gentle slopes only
on steep 'permanent' materials	relatively easy to damage
runoff slowed and filtered	runoff stopped or diverted
overtopping resisted well	overtopping causes damage
no spill structures needed	spill structures important
tend to sink into ground	erosion by rain splash rodents, and runoff,
materials may need to be	therefore need maintenance
transported to site	material on-site
lowish outlay on maintenance	relatively high maintenance
waterlogging behind line rare	some risk of waterlogging behind bund
rodents may shelter but can be uncovered	rodents dig in and may be difficult to
and no damage by burrowing	hunt or may cause failure by burrowing
little livestock damage	vulnerable to livestock damage
may be convenient to dispose of stones	labour to dig bunds
from fields into lines	

Source: various sources by the author, including some material from Dupriez and De Leener (1992, p176 – Table 289)

Wattling and *staking* (*clayonnage*) is a widely used SWC approach, consisting of wooden stakes driven into the ground along the contour, sometimes with brushwood woven between them to form wattling. Freshly cut stakes may be used which take root and quickly form contour hedges which are likely to last much longer and resist damage better. Soil and organic debris accumulate behind the stake line or hedge, quickly forming a terrace of moist soil. An ideal method for smallholders, it is widely used in Jamaica and many other steep areas (FAO, 1977).

Contour bunds or ridges (desert-step farming) may be continuous or intermittent earth-bund structures (for instance, trapezoidal bunds or triangular microcatchments) – earth banks up to 0.4 metres in height (see Figure 2.4). The bunds are designed to slow and trap runoff to improve infiltration. They are not demanding of materials but must be correctly sited and well constructed. Some form harvest runoff and can sustain agriculture in quite low rainfall environments and are discussed in Chapter 3. They are suited to regions with a reasonably predictable, regular rainfall pattern and should not be used where there is marked fluctuation, where rainfall patterns are unknown, or where slopes exceed 3 per cent.

Intermittent, rather than continuous, bunds, include: half-moon pits with a downslope bund (demi-lunes or lunettes); semicircular hoops which spill surplus runoff downslope from one to another; and *orchard terraces*. These may be constructed where terrain is rugged or levelling is difficult. In general they are of similar form to many of the water-spreading or spate-irrigation measures discussed in Chapter 4.

Contour bunds can also be used in very high rainfall environments, provided the soils are suitable (not too sandy), to enable cropping without excessive soil erosion. In the highlands of Papua New Guinea there is intensive semipermanent (semi-sedentary) cultivation, growing crops like sweet potato (*Impomoea batatus*), often on steep slopes with the aid of contour bunds (Sillitoe, 1993).

Spreader-seepage furrows are a form of water-spreading bund, similar to contour bench terraces. They are suitable for construction by farmers with access only to hand tools, and offer a means of preventing and controlling gullying. When a farmer finds a gully or a point where runoff has started to concentrate which may develop into a gully, a spreader-seepage furrow can be dug approximately 20 metres along the contour, either side of the gully (typically taking about two man days labour with a hoe). The soil from the furrow is thrown up downslope to form a bund. Provided the soil is not too stony and the ground not steep sloping, the spreader-seepage furrow should prevent or cure gullying and improve moisture infiltration along its course.

Hillside ditches may be built to hold runoff so that it can infiltrate, or to collect and divert it to a tank, cistern or 'run-on' site for agricultural use (Pereira, 1989, p158 noted that this term was once applied to deep trenches, but that approach is now obsolete, and today it is likely to mean more modest graded terrace channels). These structures are often designed to act as cut-off drains, channels which intercept potentially damaging runoff from a slope before it can destroy SWC structures. Hillside ditches slope gently along the contour to ensure manageable runoff, often with 'ties' or similar structures to help control flow (Grove, 1993) (see Figures 2.4c and 2.8).

Various indigenous strategies have evolved which essentially use hillside ditches to divert water from springs or streams, often with brushwood or stone dams. These ditches or contour channels typically have a gradient of about 1 in 100, and conduct water to terraces or hillside fields (see cover photograph). Such irrigation systems can be found in Morocco, Ethiopia, Latin America, parts of East Africa, Madeira and many other regions (Alemayehu, 1992; Leach and Mearns, 1996, p157). There is little to go wrong, no need for pumps, and a throughflow of water sufficient to prevent salinization (see Chapter 4).

Soil pits (*zay*, *zaï* or *tassia* in Mali, Burkina Faso and Niger respectively) are planting pits of various forms. Systems of *zay* are well developed in Burkina Faso, and in Tanzania there is a similar *wa matengo* pit system. They are one of the simplest forms of SWC, and possibly one of the oldest (Reij et al, 1996, p83) (see Figure 2.5a). Soil pits are a form of microcatchment formed by excavating holes (typically about 30 centimetres in diameter and 20 centimetres in depth), which are then filled with soil and compost. Sometimes the holes are arranged on the contour, sometimes in association with stone lines. They may also be dug on gentle slopes with no particular pattern. The pits trap runoff and hold moisture, silt and organic debris – for instance, leaves and animal droppings – that would otherwise be scattered and lost. Pits may have a bund on their downslope side.

In some regions termites are attracted by the compost added to the pits and their activities help to further open up the soil to infiltration. Pits are

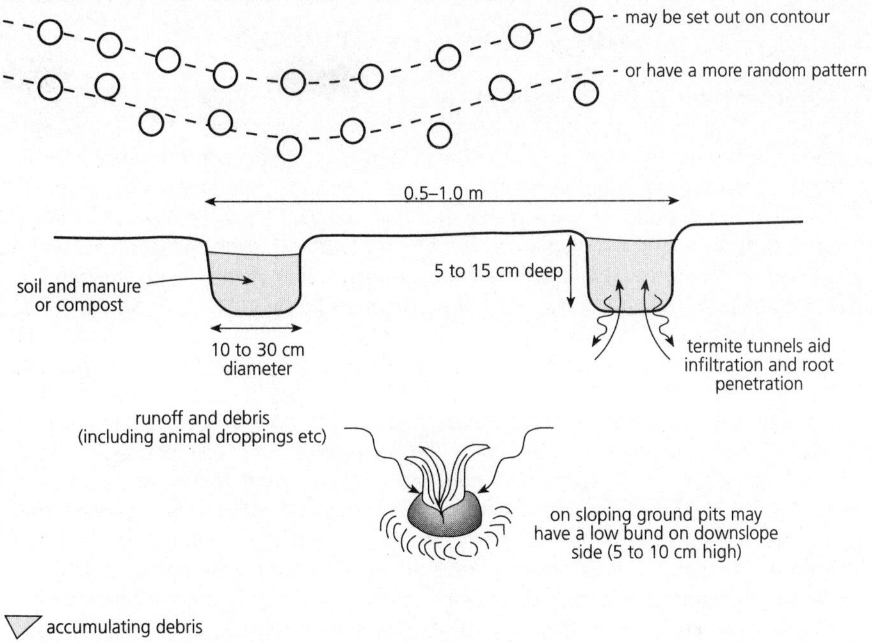

- may be set out on contour

- or have a more random pattern

0.5–1.0 m

soil and manure
or compost

5 to 15 cm deep

10 to 30 cm
diameter

termite tunnels aid
infiltration and root
penetration

runoff and debris
(including animal droppings etc)

on sloping ground pits may
have a low bund on downslope
side (5 to 10 cm high)

▽ accumulating debris

pits allow farmers to concentrate scarce soil, manure, etc

Note: effectiveness depends largely on type of soil. Need periodic redigging.

Figure 2.5 Soil pits (planting pits or *zay*)

especially useful where soils form a crust or the land has been degraded and has lost topsoil; on steep slopes under such conditions they can considerably improve crop yields or the survival chances of young trees. When hand-dug they can be demanding of labour, but are a good way to rehabilitate or crop difficult areas, or establish trees (IFAD, 1992, p24; p83; Atampugre, 1993, p46). Von Carlowitz and Wolf (1991) found 78 per cent better biomass production from trees grown with pit planting than control surface plantings in dryland environments.

Soil pits offer a number of benefits.

- They trap debris which can maintain or improve pit fertility.
- Weeding is easier than weeding whole fields.
- They are simple to construct with hand tools.
- Pits can improve food security for rural people.
- Pits can allow farmers to make best use of scarce compost or manure.
- They are useful on hard, crusted, difficult-to-dig soils and dry soils.
- Farmers can prepare pits in dry season without waiting for rain to soften fields, so there is less chance of delayed planting and reduced harvest.
- Pits concentrate runoff and hold moisture for the crop.
- Pits can be sustained where most other simple SWC techniques would suffer damage.

BOX 2.3 FORMS OF TERRACING

Bench terraces are a common 'genus' (see Figure 2.6). They can be used on slopes as gentle as 1 per cent, or on much steeper gradients (Pereira, 1989, p157, suggested they would serve for slopes of 12 per cent – 7 degrees – or steeper). The bench may be level, level with a bund on the outer edge so that it holds irrigation water, have a reverse (inward) slope, or slope gently downhill (outward slope) (Hallsworth, 1987). Most benches are constructed as close to the contour as possible, although an alternative is to construct level- or reverse-slope terraces to drain slightly along the contour to a suitable spillway (see Figure 2.8). The latter must be carefully designed, sited and constructed to avoid erosion.

Conservation bench terraces (Zingg conservation bench terrace or flat channel terrace) were developed in the US in the 1950s as a means of rainfall 'magnification' for large-scale wheat farming (Zingg and Hauser, 1959). They are level terraces with a wide interval – leaving a good deal off of the original slope between terraces to act as runoff generators for the terraces (run-on areas) (ratio of runoff : run-on areas typically 1 : 1 or 2 : 1). The runoff slopes may be grassed, planted with forage, or cultivated, and it may be possible to have crop rotation between runoff and run-on areas to help maintain soil fertility. These terraces must be constructed on quite deep soils with precise levelling and care in order to avoid burying topsoil when they are constructed. Regular maintenance with mechanized equipment is usually needed. They may also need costly drop structures at the ends of terraces to dispose of excess water. They are, therefore, largely unsuitable for smallfarmers or on steeper slopes. Conservation terraces are suitable for smallholders with access only to hand tools.

Progressive bench terraces have evolved to reduce the labour input required, in most cases at a penalty of taking some time for them to fully form. One of the best-known techniques has been developed from an indigenous Kenyan approach: the *fanya juu* (see Figure 2.6f). This is initiated by contour ditching from which soil is thrown upslope to form a terrace (Moldenhauer and Hudson, 1988, p169). This is suitable for smallholders with access only to hand tools.

Other progressive bench-terracing approaches rely on a barrier along the contour to accumulate debris carried downslope by runoff (see Figure 2.4b); gradually a step forms upslope of the barrier, which may be raised to initiate further growth. The end result is a terrace, formed with less labour than that involved in wall construction and earth moving (Hudson, 1992, p97; Smith and Price, 1994). *Controlled-erosion terraces* can be formed with check dams, which trap debris to form a level planting site.

Orchard terraces, terraces for trees, may be discontinuous chains of half-moon shaped bunds or scrapes set along the contour, rather than fully formed, continuous terraces. Orchard terraces are widely used by agroforestry. For tree planting, terraces can be much narrower than for annual crops and can be constructed on much steeper slopes – a 50-centimetre terrace might be constructed on slopes of 100 per cent (45 degrees) or more (Pereira, 1989, p163). Orchard terraces can be a good way to provide fuelwood using steep slopes when other land is in use for food and fodder production.

Terracing can take many forms and the standard of workmanship varies greatly; all forms basically fall into two categories: runoff control terraces; and terraces irrigated from streams or diverted mountain springs (see Box 2.3). Farmers often vary their terracing over short distances to suit local conditions and needs; for example, in Nepal indigenous *khet* terraces (well constructed, stream irrigated and demanding of maintenance) and *bari* terraces (more primitive, rain fed, often little more than vegetative barriers) can be found in close proximity. Vincent (1995, p23) noted: 'Terraces have proved an enduring fascination for researchers, both in the social conditions for their formation and maintenance, and in their technical construction.'

Terraces are mainly used on slopes of up to 40 per cent (in some regions, for example, on the western slopes of the Andes of Peru, on the islands of Cape Verde and Madeira, and in parts of South East Asia terraces are common on much steeper slopes). The motives for terracing vary and can include:

- conservation of moisture by retaining runoff at a point where there is enough depth of soil to store it (in the terrace);
- prevention of mass movement (soil creep, landslide, mudslide);
- reduction of rill, gully and sheetwash damage;
- make land level enough to work it or irrigate it;
- trap organic debris carried by runoff (stream-irrigated terraces benefit from fine silt in the irrigation water, and from nitrogen fixation by micro-organisms and algae on the flooded terraces) so they can be largely self-fertilizing;
- hold manure or compost applied to the cropped terrace (on sloping land it would probably wash away);
- modify local microclimate to avoid frosts;
- improve depth of soil where it is naturally thin;
- dispose of boulders and stones scattered through soil in a labour-efficient and useful way by placing the debris in lines along the contour as the land is opened for cultivation (Bronze Age stone walls in the west of the British Isles were often constructed in this fashion) (Sutton, 1989).

Reij et al (1988, p103) concluded that the fourth and fifth were often the main reason why farmers in sub-Saharan Africa built terraces.

Where soils are easily damaged, as in loess areas of northern central China, terracing is vital, even on quite gentle slopes to prevent erosion (Veek et al, 1995). In the Cape Verde islands people have engaged in labour gangs to build terraces and crescentic microcatchments to catch runoff for reafforestation and to improve agriculture. Since the mid 1970s, large areas of the islands have been planted with mesquite (*Prosopis juliflora*), a Central American desert shrub species. Established on terraces or behind microcatchments, these now provide fuelwood, fodder and have helped to reduce soil erosion.

Terraces have been used in some regions for thousands of years (Luzon in the Philippines has terraces that have remained in production for over 2000 years). Terracing was widespread in Latin America by 500 BC; during Inca rule the extent increased, much of it producing maize and other crops in climati-cally marginal areas, often at altitudes above 3600 metres (Gillet, 1987, p410; Treacy, 1989). The Andes is named after its widespread terracing – Peru alone had over one million hectares of pre-Conquest terraces, mainly irrigated from mountain streams (Sutton, 1984; Reij et al, 1988; Herve et al, 1989; Postel, 1992, p117). After the Spanish Conquest in AD 1532, social change and perhaps human epidemics led to breakdown of community organization and considerable abandonment; nevertheless, at least 250,000 hectares of pre-Hispanic terraces are still cultivated in Andean Peru alone (Mabry, 1996, p5). In Bolivia, around Lake Titicaca, terraces produce food crops at altitudes as high as 3800 to 4000 metres. In the Hindu Kush, terraces are cropped to at least 3500 metres; there are also extensive terraces in the High Atlas Mountains of Morocco and in the highlands of Ethiopia. Andean stream- or spring-irrigated terraces, together with similar forms in Java and Luzon in the Philippines and other parts of South East Asia, are probably the world's oldest sustained agriculture and can be very productive (Wheatley, 1965; Donkin, 1979; Denevan, 1986; Hastorf, 1989; Brunet, 1990; Martinez, 1990; Margraf and Voggelsberger, 1996).

It is crucial that terraces are accurately levelled to avoid runoff damage (see Chapter 6 for discussion of various appropriate technology levelling techniques); that (as for all SWC measures) runoff is accurately estimated before construction; and that care is taken to lead surplus flows of runoff to where they will not cause damage. Spacing between terraces, terrace width and vertical interval (drop or riser height) reflect soil depth and type, slope, and rainfall characteristics, and whether agriculturalists use manual labour, draft animals or are mechanized. Terrace width is not especially important where cultivation is by hand tools or if tree crops are grown, but if draft animals or tractors are used, construction must allow for a draft animal and plough, a tractor and plough, or a farmer's hand-guided rotovator, to turn. The wider the terrace and the steeper the slope, the higher and more substantial must be the walls to avoid collapse – on a 5 per cent slope terraces of five to ten metres' width are usual; where there is a 1 per cent slope 50 metres' width is possible. Terracing may not immediately appeal to agriculturalists because it takes longer to till the land; it removes some land from production; it may sometimes restrict cropping choices (although where moisture is captured it can allow crop diversification); and construction, maintenance and repair can be labour demanding.

There are various formulae to estimate vertical interval, spacing and width (see Pereira, 1989, p157). Finkel (1986, pp85–88) discusses ways to estimate ideal vertical interval and width of terraces before construction. Where the terrace width is less than the slope between terraces, farmers can harvest runoff, in effect magnifying natural rainfall (see Figures 2.6b and 2.7). To reduce labour input and risk of collapse, sloping risers are occasionally used and may be planted with a suitable crop or grass.

Terraces, and some other SWC strategies, can degenerate even after centuries of successful use if not well managed. Such failure of management may be due to epidemic, social unrest, social change (a shift from a feudal society which reduces willingness to cooperate and drives up labour costs), or to natural disaster or incompetence. Progressive failure (terraces breaking down in a domino effect) can cause severe land damage, probably worse than had they not been built (Bensalem, 1985; Ternan et al, 1996); therefore, good management is vital. Terracing can be well designed and adequately maintained and yet still fail because of activities upslope by other groups of

(a) (i) Level bench terraces with bunded edges. No runoff harvesting, but terraces often irrigated by stream or spring diversion. (ii) Detail of bench terrace construction.

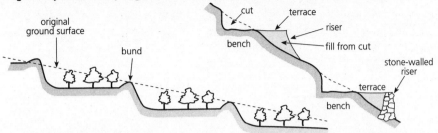

(b) Level conservation bench terraces. Runoff from slope between terraces is fed to planting strip (run-on). The wider the inter-terrace belt, the more runoff is fed to terrace; therefore, useful for lower rainfall areas. If the distance x is too great, runoff may be difficult to control. Vegetative barriers useful for controlling runoff scour and siltation.

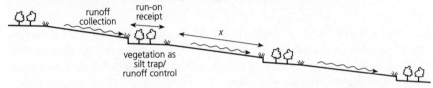

(c) Modified contour-bench terraces. Terraces slope inward to a drainage channel (toe-drain) which slopes just enough to safely feed surplus water along contour/terrace to a point where it can be disposed of with no risk of damaging downslope terraces or other structures or land (see also Figure 2.7). Sloping risers are vegetated to reduce erosion and are an easy alternative to stone walls. Suitable where there are periodic excesses of runoff.

(d) Stone-walled riser terraces. Level terrace bench formed by cut and fill. Walls must be permeable enough for adequate drainage. Care needed to ensure topsoil not buried under subsoil during construction.

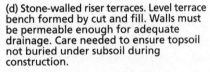

(e) Cut-and-fill bench terraces with outward sloping terraces. Suitable where there is a risk terraces may become waterlogged.

(f) *Fanya-juu* (Kenyan term). Labour-saving approach to SWC and terracing; the soil is thrown upslope to form a ridge above the furrow. Vegetation grows on the ridge and traps debris washed downslope. In time a moist, fertile 'terrace' forms (i–iii). Process may take some years. Also known as controlled erosion terraces.

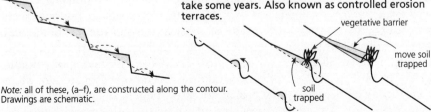

Note: all of these, (a–f), are constructed along the contour. Drawings are schematic.

Figure 2.6 Terraces

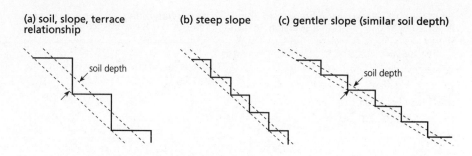

(a) soil, slope, terrace relationship   (b) steep slope   (c) gentler slope (similar soil depth)

soil depth   soil depth

Source: Moldenhauer and Hudson (1988), p123: Fig 4

***Figure 2.*7** The effect of slope and soil depth in determining maximum terrace width

people – for example, vegetation clearance, bushfire, track or road construction, all of which can collect and channel runoff causing gullying that reaches downslope to cause progressive failure below (unless SWC structures are protected with effective cut-off drains).

It is very important that terraces drain sufficiently to prevent waterlogging and have effective measures to deal with excessive runoff to avoid overtopping, leading to progressive failure and gullying. Waterlogging can damage or kill crops and may lead to salinization or acidification problems or pan formation. Ternan et al (1996) reported saturation of bench terracing in central Spain, where the 'remedy' of sloping the terraces downhill, instead of reducing waterlogging, led to overland flow and erosion. The reason seems to have been that impermeable clay layers were formed by mixing topsoil and subsoil during construction (studies by Gallart et al, 1994, give similar warnings of terracing-induced soil saturation and erosion caused by topsoil and subsoil mixing). In Rwanda, terracing was widely promoted by government extension bodies in the 1970s for runoff control, but without adequately considering how farmers would have to adjust. These terraces tended to develop acidic soil conditions, and as the farmers cannot afford lime, they dig out the upslope terrace riser by hoe and spread soil on the terrace to counter acidity and help fertilize the soil. This happens at least twice a year and therefore gradually a downslope movement of soil takes place which is greater than that which would have occurred without terracing (Lewis, 1992).

Terrace agriculture must ensure good repair – rainstorms, earthquakes and earth movement, rodents, livestock and village children can damage even well-made terraces; maintenance of soil fertility; and adequate, but not excessive, soil moisture (from direct receipt of precipitation plus runoff – or terraces may be irrigated by stream diversion). Siting is important to ensure optimum microclimate (windshelter, sunlight, frost avoidance). Terracing of volcanic soils can be especially productive. Caution needs to be exercised when constructing terraces that the subsoil is not thrown up; in addition to the afore-mentioned risk of drainage problems, terracing can bury more fertile and workable topsoil. Furthermore, the challenge of motivating agriculturalists to maintain, renew and extend terracing is often considerable.

Wherever there is a risk of strong runoff SWC must be supplied with adequate drains and spillways leading to downslope waterways. All of these may be 'tied' to form lock-and-spill channels to slow flows and to trap debris and moisture, or may be grassed (provided slope is less than 50 per cent in ideal conditions, but more likely less than 15 per cent) to slow flows and protect against scouring (see Figure 2.8). The grass used to protect waterways and channels must be sufficiently densely planted and robustly rooted to resist runoff, and may need to be protected from drought, bushfires and livestock damage, although it may still be possible to allow careful grazing or fodder collection. Steeper slopes (those between about 15 per cent and 20 per cent) require better waterway and spillway protection than can be given with grass – typically stone or tile lining is used. For slopes greater than about 35 per cent concrete or masonry channels are needed, ideally with a stepped profile (lock-and-spill drains). It may be possible to economize and use concrete, masonry or geo-textiles as drop structures just at critical points where flows change direction or level, and it can sometimes be cheaper to use prefabricated drop structures than to build them in situe.

The greater the runoff and the faster it flows, the more likelihood of channel damage and problems of safewater disposal. To design adequate spill-ways, waterways and drains for SWC it is necessary to have some estimate of maximum runoff, and rate of flow and its behaviour when channelled (this can be by empirical methods, formulae such as the rational formula, or by drawing on experience to make an informed guess). The larger the area of slope or terraces served by a drainage system the greater the runoff handled, so more small channels, rather than a few large ones, are preferable. As slope increases so does the speed of runoff up to a point – Hudson (1992) suggested that runoff from a slope of 30 per cent had probably reached maximum turbulence and erosive power, so steeper slopes pose little more challenge for runoff control.

Some terraces grade into spate irrigation methods (discussed in Chapter 4), designed to retain some soil and water and then to spill any excess as gently as possible to another terrace downslope.

Permeable rock dams are very similar to in-gully check dams (see Chapter 4); these structures are used to counter gullying but also reach out along the contour either side of the gully, so the damaging channelled flow is spread, erosion is prevented and a belt of moist soil is accumulated. A range of similar techniques can be found worldwide in India, Pakistan, Ethiopia, Mexico, Nepal and other parts of Africa and Asia. All use permeable rock barriers to trap moist silt and form planting plots (or recharge groundwater). Though the plots may be relatively small, their yield is usually well above that of much larger rain-fed fields or pasture nearby.

(a) Graded channels or cut-off drains feed into waterway via suitable drop structures.
(b) Some of the types of runoff control device that can be used to slow flow and reduce
risk of damage. Note slope (i) (>36%), (ii) (>36%), (iii) (<36%) (forms of lock-and-spill drain).
For steeper slopes (i) and (ii) continuous lining of brick, concrete or tile; for (iii) solid steps
with coarse-grass slopes between. For slopes of up to approximately 19% (depending on
soil and rainfall character, simple grassed channels may be adequate for waterways – as
in (a).

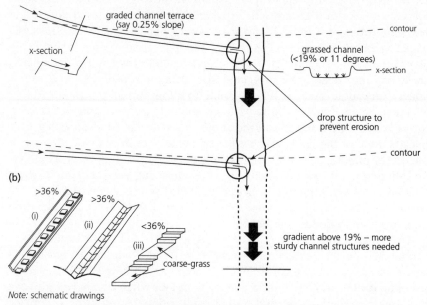

Note: schematic drawings

Figure 2.8 Disposal of excess runoff from terraces

(C J Barrow, 1978)

Photograph 2.1 Terraces at high altitude in the humid tropics –
constructed by squatter smallholders in the Cameron Highlands, Malaysia,
to grow vegetables for the Kuala Lumpur market

(C J Barrow, 1996)

Photograph 2.2 Terraces in Madeira irrigated by channels (*levadas*)
fed from springs or streams

(C J Barrow, 1998)

Photograph 2.3 Damaged terraces and overgrazed hillside in the High Atlas Mountains, Morocco

Soil and water conservation by vegetative techniques

It is important with vegetative measures that the plant cover or barrier coincides with runoff, that growth is not too slow, and that the plants provide a net saving of moisture and soil. It can take 40 days or more to grow an adequate *covercrop*, so this form of SWC needs some careful forward planning. *Strip cropping* is suitable for more gentle slopes, and consists of contour-following strips with a mix of crops, for example a rotation of cereal, legumes and vegetables to reduce downslope runoff and maintain soil fertility.

On steeper slopes it is possible to achieve SWC by means of *vegetative barriers* (vegetative buffers, live barriers or 'passive terrace formation'). These can be used alone or in conjunction with stone lines or terraces (Yadav et al, 1983; Hudson et al, 1993). Vegetative barriers may be strips of coarse-grass or herbage grown on the contour, or quickset hedges (see Figures 2.4b and 2.6f). One of the most commonly used vegetative barriers is the *contour vegetative barrier*; these are lines of robust, ideally locally adapted, plants that are planted along the contour. Like stone lines, these vegetative barriers can cheaply and effectively reduce runoff and soil loss. The trapped moisture supports a yield increase in crops, and the technique is ideal for smallholders in a wide range of environments, including steep areas and relatively dry conditions (Sharma et al, 1997). *Contour hedges* (barrier hedges) are effective in more humid environments even on quite steep slopes (15 per cent to 20 per cent) (Moyersons, 1994; Alegre and Rao, 1996). Contour hedges can be a form of agroforestry but must be carefully planned and managed if they are not to conflict with nearby crops or pasture (IFAD, 1992; Pellek, 1992; Pratap and Watson, 1994, pp39–58; Kiepe, 1995).

Agroforestry is an approach which seeks to combine vegetative barriers such as hedges for erosion control, fuelwood and fodder production with food or commodity crop production. For this to be effective it is necessary to ensure that the vegetative barrier and crops do not conflict – for instance, that they do not compete for moisture; that the barrier does not overshade crops; that it does not have an allopathic effect; that it does not harbour pests that attack crops; or that the tree or bush crop escapes to become a nuisance.

Vegetative barriers can be cheap and easy to establish compared with stone walls or terraces. As well as being accessible, they have the potential to spread fast and may yield useful products to help compensate for ground lost to erosion control structures (which could be 20 per cent of the total land area) (Manrique, 1993). Caution is needed where there are recurrent bushfires in case vegetative measures will not withstand these and be slow to recover or reestablish – during which time serious erosion may occur.

Vegetation can be used to reinforce terrace risers or contour bunds with their roots (risers are the drops between terraces). Plants such as aloe species or coarse grasses may be incorporated along the edge of terraces ('veg-edge'); or on the face of the riser, and sometimes in higher rainfall areas as a cover-crop (cover plants); and may be grown on the terrace bench between crops, and where there is a slope between terraces (see Figure 2.6c). A spiny or unpalatable plant may be selected to discourage livestock trampling, such as citronella (*Cymbopogon nadus*), aloe species, or ipil ipil (*Leucanea leucocephala*). Cover crops can improve infiltration, prevent rain-splash erosion and soil damage by sunlight, fix nitrogen, improve soil fertility, and deter unwanted weeds (Busscher et al, 1996).

It is common for growers to plant cover crops, such as creeping legumes on orchard terraces planted with oil palm, rubber, or other relatively slow growing treecrops, to protect the soil for several years until the trees mature and grow enough roots and leaves to take over, or acceptable natural weed cover replaces the planted covercrop. Covercrops, vegetative barriers and terrace reinforcement plants may provide a useful crop, such as fodder, fuelwood and cash crops of aromatic oils (eg citronella, vetiver), fruit, fibres for weaving and arabica coffee.

A common form of vegetative barrier is the *rough-grass strip*. These strips are planted along the contour (perhaps a metre in width), are cheap compared with stone terracing, require relatively little labour to establish and maintain, and over time can grow up through accumulating soil to form what in effect become terraces. The strips bind soil and filter debris and sediment, while allowing excess runoff to safely overflow, hopefully preventing gully erosion. There are grass species suitable for almost all environments and soils, from cold, semiarid regions to the humid tropics (Moldenhauer and Hudson, 1988, pp188–193; Srivastava et al, 1993, p36 provides a list of commonly used grass species). Some of these grasses are useful for cut fodder (for example, napier grass, *Pennisetum purpureum*); others such as lemon grass – a common name for several species, of which palmarosa grass (*Cymbopogon citratus*) is one – are unpalatable and resist straying livestock. Some *Cymbopogon* species produce oil that is used for perfumery and vitamin A production.

The species which has attracted particular attention since the 1980s is vetiver grass – *Vetivera zizanoides*; (in Karnataka state, India, there is quite a long tradition of using it). It is fast growing and thrives in a wide range of environments including quite dry regions, although it is best suited to areas with over 1000 millimetres of rainfall per year. Its leaves provide fodder and aromatic oil can be distilled from the roots (which demands local distilleries) (for assessments and a bibliography, see Eskine, 1988; World Bank, 1988; Anon, 1990; Grimshaw and Helfer, 1995). In spite of such promise vetiver grass can cause problems. In Haiti, people often dug up farmers' vegetative barriers to get the oil-rich roots and damaged terraces and bunds in the process.

Farmers resent 'wasting land' as vegetative barriers or terraces, but may be won round if the result is sufficiently improved crop yields, including useful products, and improved security of harvest (reduced erosion may also be welcomed but tends to have lower priority). In the humid tropics pineapples are often grown along the edge of terraces, and in the Ethiopian highlands ensete (*Ensete edulis*), and related species of false-banana, are used to bind terrace edges and provide part of the staple diet (the rhizomes and stems are used for food). Fodder crops can make useful terrace reinforcements or spill-way protection but may require livestock to be stall fed or carefully controlled. However, this is an excellent way of integrating livestock with cultivation – the manure collected from stalls can sustain terrace cropping.

In some situations it may make sense to establish more substantial vegetative barriers to provide fodder, fuelwood or timber, and sometimes to serve as wind breaks to help improve growing conditions and possibly conserve soil moisture by reducing evapotranspiration.

Intercropping can be a useful way of minimizing runoff (Vandermeer, 1992). *Alley cropping* is a generic term for agroforestry techniques which grow crops between rows of suitable shrubs or trees which can act as SWC barriers or wind breaks. Suitable shrubs include nitrogen-fixing woody legumes such as *Leucaena diversiflora*. In some countries alley cropping has been used in an attempt to upgrade or prevent the degradation of shifting cultivation.

LABOUR INPUT AND COSTS OF SOIL AND WATER CONSERVATION

Published data on SWC costs, labour inputs, and overall profitability is often vague and of dubious accuracy (Ervine, 1994; Stonehouse, 1995) (see Box 2.4). Before trying to assess costs it is important to establish the aims of the users and the constraints they face. Furthermore, there may be considerable off-site benefits from SWC that are difficult to measure and relate to on-site implementation and maintenance costs. The goal may be to get sustainable rural livelihoods through SWC; however, smallfarmers and commercial agricul-

Box 2.4 Rough Costs and Labour Requirements for Terracing

Africa

IFAD (1992, p49) lists construction requirement for selected African SWC measures (as person days): one hectare of contour terraces (Kenya) required 300 person days; 3200 square metres of trapezoidal bunds (Kenya) took 380 person days; one hectare of contour terraces on 2 per cent slope (Niger) took 90 to 110 person days. For Ethiopia, Cross (1983) reported 'typical field terraces' took 150 man days per kilometre constructed; one kilometre of 'main terrace' took 350 man days (road building, for comparison, took 200 man days per kilometre)

Peru

In southern Peru's, Colca Valley, one hectare of land supported by traditionally constructed stone-walled terraces required roughly 2000 man days (in winter) to build. This was cheaper than establishing irrigation in Peru's coastal deserts (the latter cost US$2500 to $6000 per hectare) (Treacy, 1989, p221). For the Colca Valley, Vincent, (1995, p24) found that stone walls for just under one third of a hectare of cropland, took 610 man days (1416 man days per hectare). For eastern Peru, Clark (1986) estimated that terracing for one hectare of planting plots on a 27 degree slope took 1320 labour days – far too much for a single landowner to carry out alone.

Jamaica

For Jamaica in the late 1980s: one hectare of ditches on a 36 per cent (20 degree) slope required 80 man days; one hectare of three-metre wide bench terraces on a 45 per cent (24 degree) slope required 470 man days. The annual cost of maintaining bench terraces and their waterways or spillways on 'moderate slopes' was US$175. Yields from the terraced land were about 100 per cent higher than the non-terraced equivalent; a terrace land yam crop per hectare would return US$18,750 (Moldenhauer and Hudson, 1988, pp211–213).

Note: man days and person days used – female or mixed sex labour would probably require more time than male labourers.

ture are unlikely to give much practical support to such high ideals because they are both interested in, and motivated by, short-term benefits, notably more profit, more crops, or more security of harvest (Barbier and Bishop, 1995). It is probably best to accept this and to promote SWC and runoff agriculture as a means to deliver those short-term benefits.

Without information on, for instance, the technique and slope, labour and cost data are of limited value. Construction costs should include transportation of materials to site, and labour input should indicate sex of the work team and, ideally, their state of health and the season during which the work was undertaken; this is seldom the case (the data in Box 2.4 is not consistent in these respects). Labour is often undertaken by households or village groups, the productivity and skill of which varies a great deal, or by government-funded work teams (which also differ in outputs). Reports may not distinguish between the type of group. Furthermore, building terraces and other SWC structures is not always enough; to get a deep enough planting plot may require that the farmer transports soil and compost to the terrace site, and irrigated terraces need construction and maintenance of the water supply and drainage systems. Where wet rice is grown, puddling and levelling of the terrace, and other skilled labour inputs, are required.

THE ROLE OF WOMEN IN SOIL AND WATER CONSERVATION

Often women do the majority of agricultural labour, especially where there has been male outmigration. They usually have an already heavy workload with childrearing, water and fuelwood collection, and housework, in addition to farming tasks. There is also a risk that women may carry out SWC and find the benefits taken by menfolk, so care is needed to ensure that SWC benefits are not usurped (see Chapter 6 for discussion of the role of women in upgrading runoff agriculture). Often women seek SWC for agriculture close to villages for safety and to reduce walking. It is important to ensure that construction and maintenance is manageable. As discussed earlier, some of the literature reports labour input in person days or man days; these are not standardized units, which makes it difficult to judge suitability for women.

It is desirable that women get aid and training if they are involved with SWC (IFAD, 1992, p53). Particularly in Africa, women's self-help groups have been active in adopting stone lines and terracing, and there has also been interest in garden cultivation and SWC from women in some city slums as a means of improving family livelihood (see also Chapter 7). Clearly, the role of women in SWC and runoff agriculture is considerable and could increase – it therefore deserves much more support and research.

3   RUNOFF HARVESTING AND STORAGE

This chapter considers the collection of surface runoff (overland flow), subsurface runoff flowing at shallow depth (< ten metres), the trapping of mist, cloud, fog, and dew, and drainage from roads or paved areas (but not exploitation of groundwater, other than by means of *quanats*) for agricultural use. Most of the techniques discussed in this chapter seek to catch and hold runoff, which would otherwise miss agricultural land or flow over it to 'waste', so that it can sink into the soil or be stored in a tank or cistern. Often soils with poor infiltration characteristics (where water tends to run off without soaking in) have good storage potential (if infiltration occurs they hold the moisture well). Flood or spate agriculture – the use of runoff in seasonally flowing or ephemeral flow channels (*channel flow*) – and storage of seasonal runoff in tanks are discussed in Chapter 4.

The terminology is vague and a little confused; water harvesting is widely used, as are: runoff farming; floodwater agriculture; water spreading, floodwater farming; and water ponding. Most of these terms are rather unsatisfactory, not least because they fail to indicate water source. Stormwater farming or stormwater harvesting imply that no use is made of average precipitation events, which may not be the case. Runoff harvesting, rainwater harvesting and rainfall harvesting are better generic terms (Frasier, 1984; Frasier and Myers, 1983; Gupta, 1989). The word 'harvesting' is useful in that it describes both collecting a valued resource and transferring it to where it is needed or can be stored. The collection of snow melt, fog and mist can also be included in runoff harvesting, so this is probably the best term. (Some of the literature concentrates on harvesting for domestic water supply, rather than agriculture.) The National Academy of Sciences separated rainfall harvesting and runoff agriculture, arguing that the former supported the latter: 'Once rainwater has been harvested from slopes ... it can be used for crop production... The combination is known as runoff agriculture' (1974, p23). Bruins et al (1986, p16) defined rainwater harvesting as 'farming in dry regions by means of runoff rainwater from whatever type of catchment or ephemeral stream', and suggested a division into:

- microcatchments;
- terraced *wadi* systems;
- hillside conduit systems;
- *liman* systems;
- diversion systems.

Another rough division could be into runoff harvesting for crops *before* planting, and approaches that supply runoff *after* planting. The latter is more likely in faster-draining sandy-soil areas and demands care to avoid drowning crops or washing away soil – it might also restrict the range of crops that can be grown more than does preplanting runoff agriculture.

A simple, and perhaps the best, classification of runoff harvesting is to split it into, firstly, *run-on systems* with little or no conveyance channel (the planted plot is in or adjoins the catchment – consequently, these are sometimes termed internal catchment systems), or short slope systems – such as conservation bench terraces or microcatchments. Typically these systems would each cover 100 square metres or less – in some cases under 25 square metres. Secondly, runoff harvesting can be split into *runoff harvesting with conveyance system* – channels lead harvested runoff to agricultural or forestry plots, tanks or cisterns for storing water for domestic use, livestock or irrigation outside the catchment. Hence these are sometimes called external catchment systems (for example, a hillside conduit) (see Figure 3.1a). These are also termed long slope systems by some (or macrocatchments), and each usually covers more than 100 square metres. There may sometimes be little distinction between the two systems – for example, where one terrace takes runoff from slopes above, and so is an external system, while the next down is an internal catchment (possibly with some addition from external sources), and the next down is internal plus overflow from terraces above.

Pacey and Cullis (1986, p150) suggest yet another classification: techniques of runoff harvesting involving little or no construction to generate the runoff; techniques which use constructed features to generate and collect runoff. Alternatively one might subdivide into:

- runoff harvesting where there is quite frequent light rainfall, which seeks to maximize runoff to a plot tank or cistern;
- runoff harvesting which seeks to hold runoff;
- runoff harvesting that must cope with heavy rainstorms and which, rather than induce runoff, controls it and survives heavy peak flows.

There are many traditional runoff harvesting and storage approaches, some ancient, including a diversity of modern developments. Implementation can often be incremental, carried out bit by bit over a period of time, saving labour and reducing costs (Boers and Ben-Asher, 1982; Gilbertson, 1986).

Roughly one third of the earth's land surface has insufficient precipitation to support adequate, secure, sustained rain-fed cropping, although other environmental factors should do so. Many small-scale agriculturalists try to make an adequate livelihood from these lands. In some regions human and

(a) External catchment runoff harvesting. Runoff from catchment directed by bunds and/or channels to crops/pasture/cistern/orchard some distance from the catchment. Most systems incorporate a silt trap or a vegetative barrier and/or a bund to slow flow and prevent damage to cropped plot and reduce silting. Conveyance channels may need lining and, if the ground does not slope gently, drop structures will be necessary.

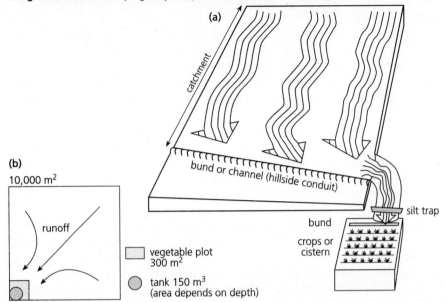

(b) The potential of runoff harvesting. For a site with an average of about 500 millimetres of precipitation per year (which could mean there are years with as little as 300 millimetres), a smallholder practising typical rain-fed subsistence agriculture needs around 10 hectares (10,000 square metres) of land to barely support an average-size family, growing millet or sorghum. If, the farmer used 1 per cent of that land (700 square metres as a catchment and 300 square metres for runoff-irrigated crops), with water storage in a 150 cubic metre tank or cistern, the family should enjoy much improved diet and security of harvest, with perhaps occasional, crop surpluses for sale. This is assuming use of a watering can to irrigate a plot of vegetables from the tank or cistern.

Source: (a) Pacey and Cullis (1986), p4: Fig 1.3; (b) Stern (1979), pp99–100

Figure 3.1 Runoff harvesting

livestock populations are increasing markedly or practices have altered, making it difficult to sustain rain-fed agriculture. Something must be done to improve rain-fed agriculture but much of the land where this is needed is too remote and too marginal for relatively expensive, high tech solutions such as mainstream irrigation development. However, runoff harvesting approaches can be effective, cheap (Rapp and Hasteen-Dahlin, 1990; Pandy, 1991; Scrimgeour and Frasier, 1991) and easy to adopt, and therefore have great potential (National Academy of Sciences, 1974, pp9–22; Shanan and Tadmore, 1979; Hutchinson et al, 1981; UNEP, 1983; Pacey and Cullis, 1986; Giraldez et al, 1988; Laryea, 1992; Gould, 1994; Kronen, 1994; Tabor, 1995; Tsioutis, 1995).

A diversity of methods may serve for supplementary irrigation (to improve yields or security of existing crops, or allow crop diversification), or provide virtually all crop, pasture, forestry or livestock needs (Suleman et al, 1995).

Runoff harvesting can also be valuable for rehabilitating degraded land (Kamra et al, 1986) and for assisting the conservation of wildlife. Despite these advantages there are many regions where people once used runoff harvesting and have abandoned, or are in the process of abandoning, it (see Chapter 5 for further discussion).

Only a portion of precipitation is shed as runoff, often less than 10 per cent of the total; the rest evaporates, infiltrates or gets caught in depressions. However, if the right sort of catchment is used, even relatively low annual precipitation can often be effectively exploited, although there are some environments where climate and geomorphology, soil character, or other factors make runoff harvesting impracticable.

Runoff harvesting is especially valuable in drylands, but it can also help better-watered regions, even for environments as humid as Amazonia which can have periods without much precipitation that stress and endanger crops. Where it is practised, runoff harvesting should achieve one or more of the following: improved security of harvest, increased yields, diversified crops, soil conservation, and improved sustainability (Nilssen-Petersen, 1982; Laryea, 1992). In drier environments, runoff harvesting, because of its 'rainfall multiplier effect', can make cropping, livestock raising or forestry possible or more secure even though rainfall is technically too low. But runoff harvesting alone may provide inadequate support for drought-sensitive crops in unusually dry periods. Where droughts are a risk, runoff harvesting should be combined with water storage (as moist soil, or as water in cisterns or tanks – if economics, geology and terrain allow) (Verma and Sarma, 1990). For example, a 1.2 hectare catchment in the Negev, where there may only be 100 millimetres of rain per year, can supply a 440 cubic metre cistern which could reliably support 300 to 500 sheep (Pacey and Cullis, 1986, p27).

With global warming a distinct possibility, runoff harvesting may become necessary in areas where precipitation is currently adequate for rain-fed farming. Climatic change may also make runoff harvesting possible in areas where it is today not really viable, and perhaps make it less practical in others. Proposals to develop runoff harvesting must therefore consider future climate change predictions, although forecasting is imprecise. The need for runoff harvesting is also likely to be driven by increasing demands for available water resources as a consequence of population growth, urban and industrial expansion and modernization (Mou Haisheng, 1995). Runoff harvesting has attracted particular attention from those concerned with the need to improve agricultural production in sub-Saharan Africa (Critchley et al, 1992; Cullis and Pacey, 1992; Siegert, 1995; Tabor, 1995).

Before any decision to promote runoff harvesting, the planners must try to obtain good data, especially on the distribution of precipitation with respect to growing season. Reij et al (1988, p27) cautioned that it is better to check the number of rainfall events likely to yield runoff during the year than to depend on mean precipitation figures (Cohen et al, 1995). For an examination of risk assessment applied to runoff harvesting, see Cohen et al (1995). Modelling of runoff has generated a large literature, some of it focused on runoff harvesting for agriculture (Oron et al, 1989; Sorman et al, 1990; Rees et al, 1991). There are ongoing studies to explore the potential of integrating prediction and optimal use of rainwater. These use combined geographic information systems (GIS) and water harvesting, storage and utilization modelling computer programmes (*IDRC Review of Research for Development*, 1997–1998 issue, pp34–36).

When runoff harvesting is possible, but moisture storage measures are not, it may still allow farmers to get better crops some years and so accumulate reserves of cash or food 'buffer stocks' for drought or famine periods. Tabor (1995) noted 'spectacular' results with runoff harvesting in the Sahel zone of Niger and potential for far more (see also Tauer et al, 1991). Runoff harvesting is of considerable importance already, and has even greater future promise for improving world food production, especially in marginal, often remote areas, for improving forest cover, and as a means of rehabilitating degraded land (Le Houérou and Lundholm, 1976; Hogg, 1988; Reij et al, 1988).

Archaeological work in the 1950s, especially in the Negev Desert of Israel, helped to trigger recent interest in runoff harvesting and microcatchments. Between the Bronze Age (approximately 2000 BC) and roughly the 7th century AD the Negev supported quite a large rural population, partly related to important trade routes which ran through the region. By around AD 450 agriculture in the region fed the cities of Petra, Avdat and Shivta, in spite of a harsh, low rainfall environment. Studies of the ancient Nabatean people's runoff harvesting systems in the Negev Desert, some more than 2000 years old, prompted the establishment of several research farms between 1958 and 1967, notably those at Avdat and Shivta (Evenari et al, 1971; Evenari et al, 1982; Adato and Miller, 1986; Nevo, 1991; Johnson and Lewis, 1995, pp29–41). These farms extended archaeological study and have been testing and developing the ancient techniques of *wadi* diversion agriculture and water harvesting by means of internal and external catchments. Some of these farms have catchment-to-planted area ratios of 30:1, which enables cropping or livestock rearing with annual (mainly in winter) rainfalls averaging less than 200 millimetres per year (often less than 100 to 150 millimetres per year). The techniques seem to offer great potential for other similar drylands and have been adopted with some success in parts of Afghanistan, Australia and Africa (Shanan et al, 1969; Yair, 1983; Bruins et al, 1986).

Nevertheless, the Negev experience may be difficult to transfer to some areas because it is a region fortunate in having a growing season that coincides with winter rain, together with particularly suitable slopes and soils. Other parts of the world may have equal or greater annual precipitation, but it falls in the form of a few intense summer storms when temperatures are high and evapotranspiration losses are considerable, or soils less readily shed runoff for collection or contaminate it with salts (Robineau and Robineau, 1988).

Archaeological and anthropological studies of runoff agriculture traditions in Latin America have also stimulated interest, shown the potential of traditional methods, and helped to develop modern approaches (Rohn, 1963). India also has a long tradition of runoff agriculture and has carried out research and development in the field since the 1970s or earlier. In the southern US, in Colorado, Utah, South Arizona, New Mexico and the Mojave Desert, the

Navajo, Hopi, Papago, Zuni, Mojave, Yuma, Cocopu and Maricopa peoples have long traditions of runoff harvesting in harsh environments (Nabhan, 1983; 1984). Studies in the southern Yemen have also helped stimulate interest and prompt development and extension of techniques (UNEP, 1983, pp140–145; Brunner and Haefner, 1986; Bentley, 1987). There are also valuable traditions of runoff harvesting agriculture in Africa, notably Morocco, Tunisia and Libya (Kutsch, 1982; 1983; Finkel, 1986).

Techniques based on Negev experiments were spread to Upper Volta, Mali, Niger and Kenya (Hillman, 1980). However, there are considerable areas which have conditions which should allow the adoption of Negev-type approaches – for example, large areas of Nigeria have winter rainfall and soils that are probably suitable.

RUNOFF HARVESTING TECHNIQUES

The rainfall multiplier effect of runoff harvesting depends primarily on the ratio of catchment area to run-on area (planted area), although there are losses of moisture to seepage and evaporation that vary from site to site. Some soils are better for catchments than others; for example, vertisols are not as good as alfisols at shedding water (and it may be necessary to treat the catchments on the former to reduce infiltration losses). Runoff harvesting, with a planting area within catchment, usually has a catchment:run-on ratio of between 1:1 and 5:1, while runoff harvesting with an external run-on area often has ratios of 20:1 or greater. The yield from runoff harvesting can be predicted with empirical formulae (Ben-Asher et al, 1985; Hudson, 1987, p50) or water balance model approaches (Boers et al, 1986a; Sanchez Cohen et al, 1997). Catchment size is a significant factor, with smaller catchments tending to be more efficient in yield and providing runoff with less delay (see Figure 3.1b).

The plot in receipt of runoff usually needs appropriate preparation; where soil is thin a pit may be dug and filled with soil and compost to hold moisture around the crop or tree roots. Where soils crust over it may be necessary to till before receipts of runoff. After infiltration it may be worth mulching a plot to help reduce evapotranspiration.

There are four ways of storing runoff: excavated tanks; underground cisterns; small dams; and as soil moisture (Bateman, 1974; Pacey and Cullis, 1996, p132). A cropped area must not remain under water or with the root zone waterlogged (saturated) for more than a short while since this will kill the plants. The maximum amount of moisture a soil can hold without being waterlogged is known as *field capacity* (see Glossary). If runoff is irregular, especially if crops are to be grown in summer, soil at field capacity may not provide enough moisture to keep the crop flourishing until the next rains. A solution is to saturate a plot (exceed field capacity) and plant when it has drained enough for the upper few tens of centimetres to have drained to field capacity. The plant roots may then 'pursue' the falling watertable as they grow, extending crop production until more runoff is received (this is the way techniques such as the *khadin* or *ahar* operate – see later in this chapter).

Runoff agriculture must balance effective collection of moisture (which means ensuring a flow of runoff) and prevention of excessive soil erosion. Runoff may carry quite a heavy load of debris, so effective silt traps may be needed or planting plots and cultivation practices should be used that are able to cope with debris-laden runoff. Some runoff agriculture techniques capture silt and debris to form a reservoir of moist sediment or add organic matter swept from the catchment to the planting plot in order to improve fertility. This fertility enhancement makes some irrigated terrace and planting pit-type systems sustainable for centuries with little or no inputs (other than labour and water). Where runoff harvesting is intended to store water in tanks, dams or cisterns, channel it some distance, or irrigate terraces, it is important to install silt traps.

Roaded catchments

Roaded catchments were developed in Western Australia by commercial farmers in the 1950s because their land suffered from poor precipitation and there was little available groundwater (Reij et al, 1988, p23; Pacey and Cullis, 1986). These cambered catchments are mechanically compacted and runoff down channels between the 'roads', either to tanks from which it can be drawn to water livestock or crop land, or to moisten a planting strip between the cambers (see Figure 3.2a). They can be undertaken on level or gently sloping areas where soil is favourable.

There are limits to the slopes on which roaded catchments can be formed – Suleman et al (1995) report problems with rill erosion of compacted soil catchment surfaces where the gradient exceeded 7 per cent to 15 per cent (depending on the type of soil).

There have been studies exploring the possibility of combining control of farmland soil compaction by agricultural machinery with runoff collection; compacted routes would support specially designed farm vehicles that cultivate and harvest, and with regular compaction would be better able to gather runoff (Spoor and Berry, 1990).

Enhanced catchments

To some extent any modified natural catchment is artificial, particularly when the soil has been moved and compacted, as with roaded catchments. However, the term is usually applied to catchments treated with some form of sealant. Catchments constructed of concrete, or plastic, rubber or metal sheeting are used by a number of islands for domestic supply and livestock. In Bermuda virtually every household collects rainwater from roofs for domestic and horticulture use; Gibraltar has extensive corrugated-iron sheet catchments for domestic supply (one over 14 hectares in area). Not all soils shed rainwater readily. The quantity of precipitation required to yield runoff is known as the *threshold precipitation*; this varies with terrain, vegetation and soil

(a) roaded catchment (can be formed with tractor and modified plough; popular in western Australia)

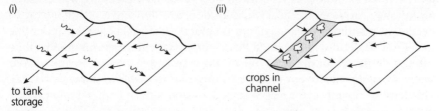

(b) within-field catchments (bunds control runoff and are often a good way to dispose of loose stones on catchment)

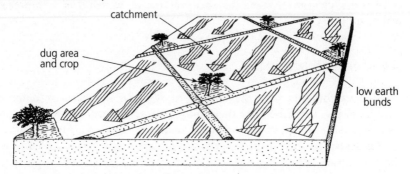

(c) contour-strip catchments (feed runoff to cropped strip, often terrace – strips run along contour)

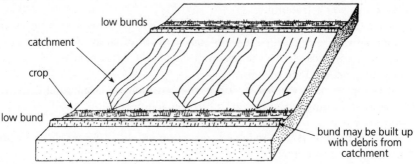

Note: diagrams are schematic and not to scale (slopes are much exaggerated)

Source (a) author; (b) Pacey and Cullis (1986), p10: Fig 1.4b; (c) Pacey and Cullis (1986), p10: Fig 1.4a.

Figure 3.2 Types of runoff catchment or microcatchment

characteristics. A locality may have quite a reasonable annual precipitation but it may fall as frequent light showers that do not exceed the threshold precipitation and so evaporate or soak away. There may be situations where catchment treatment will reduce infiltration and make the surface water repellent enough to exceed the threshold precipitation or boost yield where collection is viable.

The problem is to find and effectively apply a cheap, stable and non-harmful compound; concrete, metal sheets, and rubber sheets are used by commercial agriculture but are too expensive for smallholders (UNEP, 1983, pp49–61; Barrow, 1987, p176). Precipitation-collecting areas (catchments) may be enhanced by being cleared of debris and vegetation and perhaps smoothed to prevent puddles which increase evaporation and seepage losses (Lindstrom, 1986). In addition to using plastic, rubber or metal sheets, catchments can be enhanced by artificial compaction; chemical treatment to cut infiltration, such as with sodium salts (NaCl), gypsum compounds or sodium methylsilanclate to seal clays (Frasier et al, 1987); sprayed with dispersants to better shed water (for example, waxes or silicone compounds); managed to encourage a sealing algae layer; covered with paraffin wax (Fink, 1984), sprayed with oil, latex, tar or plastic emulsion; or coated with hot bitumen (Hutchinson et al, 1981, p17; Bohra and Issac, 1987; Benhur, 1991). Arnon (1972, p135) reported that applications of common salt (at 45 kilogrammes per hectare) improved runoff from clay–loam soil by 70 per cent. This is a cheap treatment but demands care to ensure that there is no long-term risk of soil salinization. Microcatchments are less likely to be sealed, other than by compaction, than larger catchments which feed external run-on plots, tanks or cisterns.

The enhancement of a catchment must be cost effective and safe; each treatment varies in effectiveness with local soil and other conditions, and there is a considerable range of performance between treatments (Emmerich et al, 1987). The best runoff yields, without resorting to chemical treatments or artificial sealants, is achieved by removing loose stones and debris, clearing vegetation, preventing puddle formation, and if need be compacting the surface.

It is unlikely that these sorts of catchments could be of much use for smallholder agriculture; however, highways and school yards can provide irrigation water for horticulture, and in some countries people have opportunistically made use of such supplies. In developed countries runoff from such sources could be used deliberately to form wetlands or establish vegetation for wildlife conservation.

Microcatchments

Microcatchments (minicatchments may sometimes be used for particularly small microcatchments) are run-on systems that can be constructed on flat, gently sloping and steeper slopes (certainly up to 3 per cent). They can provide water for crops (Renner and Frasier, 1995a; 1995b), forestry or conservation of flora and fauna. Each microcatchment feeds a single tree or shrub or a small cropped or grazed area. Microcatchments are usually cheap and simple to construct, involving little more than the use of hand tools to scrape and build low bunds, although, as in North America or Australia, they can be constructed by large scale farming with modern mechanical aids (Tsakiris, 1991) (see Figures 3.2b; 3.2c; 3.3a,b,c and 3.4). They are often useful for smallholders in harsh environments and for tree planting and pasture improvement (Critchley, 1990). They can also be used for land rehabilitation, especially where funds are in short supply, and are a valuable way of establishing shelter belts and fuelwood plantations.

(a) Simple microcatchment. Can be used on gently sloping (i, ii) or level land (iii, iv). Planting pit is dug to at least 1.5 metres depth and the soil is enriched with compost or manure. *Source:* Redrawn from a number of sources,including: Evenari (1968); Reij, et al. (1988), p42: Fig 4.

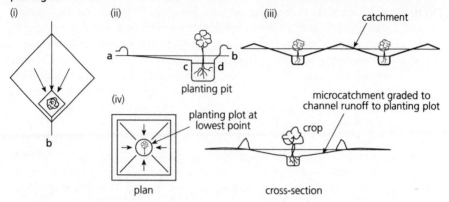

Note: arrows indicate direction of runoff flow. Cultivated plot is placed at the lowest point of the natural terrain within the catchment; its position varies. Walls are 15–20 cm high; pit surface is about 40 cm below the catchment, holding water close to the plant; root-zone soil must be at least 1.5 m deep; the microcatchment can be less than 5 m or more than 30 m across, depending on climate and crop.

(b) Contour-ridge microcatchment (*negarim* type). Suitable for gentle to moderate slopes. Developed in Negev Desert of Israel. *Source:* sketch based on several sources.

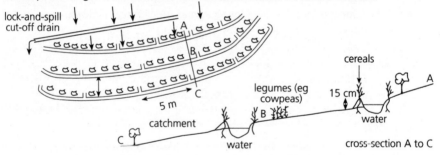

(c) Contour bunds. Semi-circular type (ii) often termed half-moon or demilunettes or lunettes. Trapezoidal bunds must usually have seepage drains or spillways to prevent overtopping and damage. The catchments are likely to be bigger on gentle slopes, and type (ii) and (iii) can be much more than 12 metres across and may be excavated to hold more runoff (for related structures see Figure 3.6: *teras, haffirs, mahafurs*).

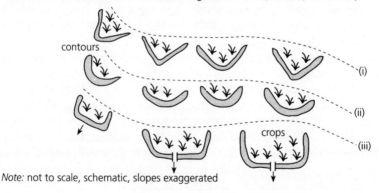

Note: not to scale, schematic, slopes exaggerated

Figure 3.3 Microcatchments

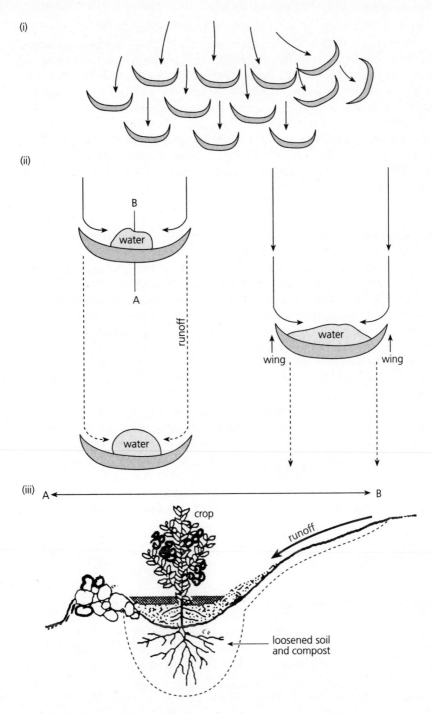

(i)

(ii)

(iii)

Note: i and ii = plan; iii = x-section.

Figure 3.4 Microcatchments: half-moon type

Microcatchments have great potential for a wide range of environments, ranging from drylands to more humid areas (National Academy of Sciences, 1974, pp29–37; Oweis and Taimeh, 1977). The soil and slope must be suited to the rainfall distribution and intensity, however, or there will be difficulties. Problems arise where there are a few very heavy storms and mainly summer rainfall, or where precipitation is unpredictable and very erratic. In addition to Israeli development of microcatchments in the Negev, there has been considerable research and development in India (notably by the Central Arid Zone Research Institute, Jodhpur–Rajasthan), south-western and southern US (especially in Texas) and in Australia. The techniques have spread to Botswana, Niger and Kenya.

Small catchments tend to offer better runoff efficiency (waste less water) and make it less of a challenge to control erosion because flow volume and speed are limited and control structures are more frequent than is usually the case with large catchments (Sharma, 1986). In terms of function, terraces often operate as microcatchments (see Figure 3.2c), but 'microcatchment' is usually applied to earth bunds or excavated features which channel runoff.

There is relatively little conveyance loss with microcatchments but a rather low cropping or planting density. Under natural conditions, perhaps as little as 5 per cent of rainfall will reach a stream as a consequence of interception, evaporation and infiltration; however, a microcatchment can yield as much as 50 per cent of precipitation to the cropped area or tree (Pacey and Cullis, 1986, p7).

Microcatchments are widely used for tree planting (Barrow, 1983; Sharma et al, 1986), often in quite low rainfall regions. For example, in the Cape Verde Islands, annual rainfall can vary between 100 and 900 millimetres per year and soils easily erode during storms where vegetation is disturbed; without microcatchments or contour ditches, tree planting would largely fail and topsoil would be carried away (Sandys-Winsch and Harris, 1994). An Israeli–German study in the Negev proved that minicatchments can be used to establish trees in suitable drylands with as little as 34 to 167 millimetres rainfall per year (Tenbergen et al, 1995). Gupta (1994; 1995) examined the effectiveness of microcatchments for establishing trees in the Indian Desert and concluded that it had potential.

Macrocatchments

Macrocatchments (composite catchments) collect enough water to necessitate substantial walls and runoff-control structures (see Figures 3.5 and 3.6). The complexity of these measures, as well as the scale of construction, tends to make macrocatchments less suitable for individual smallfarmers. Usually such measures require cooperative inputs of labour. Many relatively large-scale indigenous runoff harvesting techniques collect natural runoff (there is little or no construction to generate the runoff – so 'macrocatchment' is possibly misapplied since the catchment is not improved upon). These systems collect or divert considerable amounts of runoff and often involve substantial bunding, channel structures and perhaps tanks or cisterns (FAO, 1988, p138).

One of the simplest forms of macrocatchment is the *kutsch* technique of Morocco, which involves little more than improving a natural hollow where runoff is expected to accumulate to allow cropping as the watertable falls. However, in practice to do this effectively requires considerable skill and expertise. Another ancient form of macrocatchment, the *gessour* (*jessour*), is used in Tunisia and the Yemen (in areas of 100 to 150 millimetres rainfall per year) for wheat, olive and date growing. A typical *jessour* has a catchment:planted plot ratio of 2:1 (Gischler, 1979, p53–56). *Jessours* can also make use of diverted runoff or seasonal stream or river floodwaters. In Tunisia, the *meskat* system is still in widespread use, although there has been some break down in recent decades. *Meskats* allow viable agricultural production where there is around 300 millimetres rainfall per year, and demand only local materials for construction (Bonvallot, 1986) (Figures 3.5a and 3.5b). Because *meskats* are quite demanding of labour, in many places they have been abandoned or are in decline as a result of rural–urban migration (El Amami, 1977). A good proportion of the olives and grain for ancient Rome were produced by *meskat*-type methods, mainly in Libya (El Kassas, 1979; Barker and Jones, 1982; Gilbertson, 1986; Gilbertson and Chisholm, 1996; Gilbertson and Hunt, 1996).

These types of runoff agriculture could be used more widely. Similar types of system are used in Botswana (having been introduced) and have catchment:cultivation ratios of between 17:1 and 50:1. Carter and Miller (1991) found that these give two or three times the yield of nearby rain-fed cultivation. Experiments in the Negev (Israel) with *negarim*-type catchments (structures based on indigenous methods of the Negev Desert) gave much better yields than established rain-fed cultivation: 2000 kilogrammes per hectare compared with 45 kilogrammes per hectare (Lewis, 1994).

Liman systems

Liman (*limanim* plural) (or *wadi*-floor bunds), and the *teras* of Sudan, east of the Nile; Uttar Pradesh; and Bihar (India) are broadly similar. *Limanim* spread *wadi* (ephemeral stream) flows across valley terraces to absorb water and trap silt (see Figure 3.6), or to trap moisture and soil, usually behind quite large earth bunds which are built to intercept runoff and fine sediment from slopes or streams. A number of variants have been developed to exploit natural overland flow from gentle slopes. Most consist of cross-slope bunds, often of crescentic form, or semi-circular excavations in the land surface, or a combination of both.

Some of these *liman*-type systems collect water for livestock and some trap water and silt for crop growing. Figures 3.6a and 3.6b illustrate two traditional forms of *liman*-type systems: the *mahafur* (plural *mahafir*) (an ancient earthen-bund runoff harvesting technique found in Jordan), and the *teras* (Agnew et al, 1989; Agnew et al, 1995; Van Dijk, 1997a). *Teras* are an indigenous runoff harvesting method in east Sudan, where the rainfall ranges between 200 and 500 millimetres per year, and there are clay-rich vertisols

(a) Composite type of catchment. A contour bund *negarim* type (external catchment). Spreading in sub-Saharan Africa (eg Kenya – see, Lewis, 1984). Suitable for slopes of up to 3%. May be constructed by hand. Catchment to cropped area ratio depends on rainfall, but is typically about 5:1 for a dryland region.

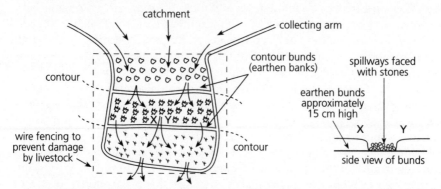

(b) Traditional composite catchment (internal catchment). Tunisia and some other Maghreb countries developed these *meskat*-type (or *impluvium* catchment) centuries ago. The cropped area in receipt of the runoff (the *mankaa*) may produce olives, dates, and perhaps some grain and other annual crops. Widely used for olives where there is 200 to 400 millimetres of precipitation per year. Large numbers established in Tunisia by Roman times, many abandoned but still important in some areas. Typically the catchment to planting area ratio is about 2:1. Suitable for vertisols and slope of 0.4 to 0.8 per cent.

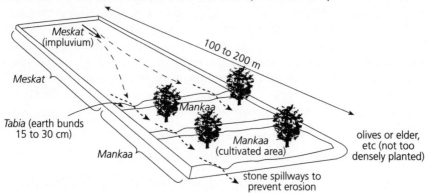

(c) Larger macrocatchment structures (called *khadin* or *ahar* in India). Extensive earthen bunds trap water and silt. Crops planted as trapped water seeps away, and may also be irrigated from wells dug into the moist ground.

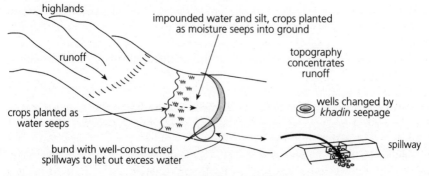

Source: redrawn from various sources including Reij et al (1988) p60: Fig 10.

Figure 3.5 Macrocatchments

(a) *Mahafur* (plural *mahafir*) runoff collecting structure. Excavated into gently sloping soil (playa surface), with debris used to make a crescentic bund. Stores a pool of water/wet soil which is mainly used for livestock watering (E Jordan), sometimes cropped. Similar to *haffir* (see Bryan, 1929; Agnew et al, 1995)

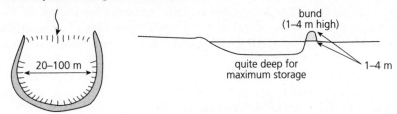

(b) *Tera* (two forms of bund). Similar in construction to (a), but not excavated so much or at all. Trap runoff and organic debris to maintain fertility for cropping. Common on vertisols (clayey soils) in east Sudan, where each provides 0.2 to 3.0 hectares of cropland. Decline in 1970s; some revival with adoption of tractors to construct (see, Van Dijk, 1977; Van Dijk and Ahmad, 1993).

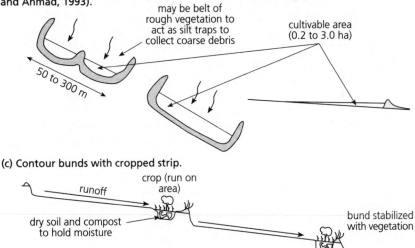

(c) Contour bunds with cropped strip.

Figure 3.6 Runoff collection

which resist infiltration. Unlike the *mahafir* or *haffir* (see later in this chapter), they are simply bunds *without* an excavation to help store water and silt (Van Dijk and Ahmed, 1993).

Liman-type systems are often difficult to separate from spate irrigation techniques (see Chapter 4); the *khadin* (found in north India, especially the Thar Desert of west Rajasthan), and the *mahafir* can be regarded as runoff harvesting when they collect overland flows, as opposed to ephemeral stream-flow in channels (Reij et al, 1988; Mohammed, 1993; Agnew, at al, 1995). The *khadin* techniques were developed long ago around the Thar Desert of western Rajasthan (a similar approach in Bihar is termed an *ahar*) (see Figure 3.5c). *Khadin* are common in parts of India with deep soils and rainfall of less than 500 millimetres per year (Stiles, 1995, p200). Over time these *khadin* accumulate fine clay–loam as well as moisture, offering a much better planting medium than surrounding soils (Pacey and Cullis, 1986, p140). Some Indian *khadin* can have bunds over three metres high and can pond several square kilometres (Hudson, 1992, p20, reports one of 19 kilometres in length impounding 41,000 hectares); they are especially useful for storing seasonal monsoon rainfall so that a second crop (*rabi*) can be grown during the following dry winter (Hudson, 1992, p137). *Khadin* have a rock-built spillway to release excess water so that the earth bunds are not overtopped and damaged. *Ahar* can be as much as 4000 hectares in area (UNEP, 1983, p164). There are similar systems in Latin America; for example, in north-eastern Brazil earthen bunds are used to hold runoff on *várzantes* – areas where there is sheet runoff during rains but poor infiltration into the soil.

Teras can have bunds ten kilometres long, impounding 4000 hectares or more. They are found in the Sudan, east of the Nile; North and South Yemen; and South Pakistan – where they are called *sailabas* or *kuskabas* (they are called *khadins* in Rajasthan and *ahars* in Bihar). *Teras* are cropped once the water and debris they have trapped in the wet season sink enough for the crops to get established; *teras* then 'pursue' the sinking moisture for their approximately 80-day growing season (Pacey and Cullis, 1986, p131). The size of these structures varies a great deal; some can run for several kilometres. Usually one structure spills excess runoff to another downslope, and so on (Kolarkar et al, 1983). Included in this category are water-spreading techniques; these are a diverse group of techniques often evolved in areas with poor infiltration but reasonable storage-capacity soils (typically clays that infiltrate slowly but which retain moisture well) (see Figure 3.6). Where soils are sandy it may be necessary to wait for some years after construction or repairs for fine silt and algae to seal the ground enough to hold sufficient moisture. The main problem with these structures is the risk of overtopping and breaching when there are severe floods (Van Dijke, 1997b).

Niemeijer (1998) reported that *teras* in east Sudan yielded the equivalent of 650 kilogrammes per hectare of grain compared with rain-fed yields of between 150 and 250 kilogrammes per hectare. The *teras* also seemed to be more sustainable and had less need of chemical fertilizers. Therefore, the value of *teras* is as much that they harvest fertile debris as conserve water. *Haffirs* are excavations made in natural catchments to collect runoff for livestock watering and sometimes cropping. They are found in south Sudan, east Sudan, Ethiopia and Kenya, and may store as much as 200,000 cubic metres of water (UNEP, 1983, p158; p219). These structures appear to require at least 80 millimetres per year of precipitation during a cold, rainy season – much more if rains are during hotter weather, or if the soil is a clay or a soil which crusts, and low levels of salt are in the soil. When cropped, planting takes place as stored water seeps away – a sort of flood recession agriculture. Where tractors or earthmoving equipment are unavailable, the construction and maintenance of *haffirs* demands cooperative effort to marshal enough labour.

Terraced *wadi* systems

These too may be difficult to separate from flood or spate systems (see Chapter 4). *Wadi* diversion structures may take the form of, firstly, earth banks with spillways for excess flows or semi-permeable barriers (stone check dams, brushwood barrages or stone-filled *gabbions*) built across the *wadi* to divert flows onto flood terraces where soil can be built up for cropping; or, secondly, headworks or channels that fill only when a flood spreads away from the main channel and is flowing slowly, the water then follows channels that supply terraces or planting plots a safe distance from the *wadi*. *Wadi* diversion structures also harvest ephemeral flows and flash floods and are also considered in Chapter 4 as flood or spate agriculture. In the past some of these structures have been on a grand scale – for example, the great dam across Wadi Dhana in North Yemen (Brunner and Haefner, 1986). Some *wadi* diversion systems trap debris as well as floodwater and so can maintain soil fertility, even accumulating suitable soil where there was none before. These systems can also help to control damaging flash flooding and to sustain biodiversity by improving moisture and detritus supplies to floodland vegetation and wildlife. Similar approaches can be applied to gullied areas or ravines (Bhushan et al, 1992).

Warping

Warping strategies which seek to trap and accumulate silt and organic matter to form moist, fertile soil fall into two categories: check dams across channels to accumulate plots of a limited extent; and diversion of floods onto almost-level river floodplains to flood the land and deposit (hopefully fertile) silt over quite large areas (Reij et al, 1988, p59). Warping is an ancient technique, for example, widely practised along some rivers in the People's Republic of China, especially those draining the north-central loess soil plateaux (UNEP, 1983, p174). *Water meadows*, once common in England, are also a form of warping (some are natural, some engineered), although in the past they also sought to speed spring growth of pasture by flooding, reducing the effect of frost on the land. There is much in common between warping and techniques that form sand-filled reservoirs. Various forms of cultivated sand-filled reservoirs (or inundation bunds) have been developed; some are water harvesting and storage strategies, and others flood and spate irrigation strategies (see Chapter 4).

Hillside conduit systems

These structures intercept overland flow by ditch or bund, and are dug along the contour but with just enough slope to safely convey the water to agricultural plots, terraces or tanks. They are suited to situations where planting is removed from the runoff catchment – for example, where soil has accumulated in valley bottoms or as localized alluvial fans. These are long-slope, external catchment systems and may be used to harvest runoff from low-intensity rains which might not reach the valley bottom were it not intercepted and channelled. The land watered by these conduits may be terraced or bunded to retain the flow and protect against erosion; the conduits should also have strong spillways to shed excess flows and silt traps to collect unwanted debris. In some highland areas conduits collect water from streams and springs and lead it to where cultivation is possible. A spectacular example (*levadas*) can be seen on Madeira, some several kilometres long and traversing very rugged terrain.

FOG AND MIST HARVESTING

Natural vegetation in some regions traps enough passing cloud, fog, mist, and drizzle to flourish and even recharge groundwater and springs (raindrops are approximately 0.5 to 5.0 millimetres in diameter; drizzle 1.0 to 40 micrometres; mist 40 to 500 micrometres). Removal of extensive areas of such plants may have serious consequences on water resources – as has been discovered in Tenerife, the Sultanate of Oman, and parts of southern Ecuador, Peru and Chile as far south as approximately 35° South. In Peru, coastal fogs and mists (*caman-chaca*) support the *lomas* vegetation association if at least some plants are large enough to trap sufficient moisture. Established *lomas* vegetation can trap as much as 1.8 litres per hour with a wind speed of only 1 metres per second, and misty days occur on average 120 days per year (Walter, 1971, p375). Once overgrazed or cleared, regrowth is too low to trap moisture and reestablishment is difficult (Goudie and Wilkinson, 1977, p66). Reafforestation with casurina (*Casurina* species) near Lima, Peru, has shown that once big enough they can trap enough moisture to flourish. In Hawaii, Norfolk Island pines (*Araucaria heterophylla*) have also proved effective at trapping mist and fog.

There is a tradition of collecting mist blown by onshore winds in the Darfur uplands of western Muscat and Oman (Gischler, 1979, p76), but in the main artificial collection is a more recent development. Artificial fog traps have proved capable of establishing trees which then shed enough moisture for natural reseeding. Traps can certainly provide enough water for household domestic supplies, for establishing trees and for limited livestock rearing, and in especially favourable localities may support small-scale horticulture (Pinche-Laurre, 1966; Gischler, 1982).

There have been promising trials with textile nets or plastic-sheet traps in parts of Chile to provide village water supplies and to establish treecover (Schemenauer and Cereceda, 1997, report 150 to 750 litres per day from 48-square metre polypropylene mesh traps), as well as in Peru, Cape Verde, Oman and Israel, and there are other promising areas where airborne moisture that could be trapped passes across drylands (Gindel, 1965; Schemenauer and Cereceda, 1994). Experiments in Peru with simple netting traps of 6000 square metres in area collected an average of 50 cubic metres per day; in the Negev, polyethelene-sheet traps yielded up to 3631 millimetres per square metre per month – enough to support trees (Gindel, 1965). In Chile, experiments at El Tofo showed a textile-mesh trap of 90 square metres could collect one cubic metre per day. Similar traps at Pasamayo and Antofagusta (Peru) gave 1 to 15 litres per square metre of trap (Gischler and Jáuregui, 1984, p14) – in a region which had about 5 millimetres of rainfall per year. Schemenauer and Cereceda (1991; 1992) reported average yields of 7200 litres per day from 48 square metres polypropylene-mesh fog collectors during three drought years in a coastal area of northern Chile. The authors go on to identify promising areas in over 20 countries where the relatively simple and cheap approach could be used. There may be potential in Cape Province (South Africa); Namibia; Cape Verde; Western Australia; Hawaii; the Canary Islands; the Galapagos; South Oman; and Baja California. Juvic et al (1995) describe monofilament screen fog traps in the mountains of Cape Verde which, in the right situation, captured up to 7.8 litres per square metre per day in the 1988 *dry* season. UNEP (1983, p191) claims that traps can yield the equivalent of 3200 millimetres of precipitation per year. Schemenauer and Cereceda (1997, p18) estimate the cost of a 48-square metre polypropylene mesh trap as US$400 (including construction costs, channels, pipes and cistern), and suggest it would have at least a ten-year lifetime in mountain areas where there is high UV damage.

A horticulture strategy which might serve smallfarmers in remote areas that have frequent mist and fog, but little rain, could be based on simple textile traps feeding a storage cistern, from which water is drawn in watering cans to fill unglazed earthenware pitchers (*kuzeh* or pot irrigation) that are placed close to the roots of vines, treecrops, or crop plants (Mondal, 1974). The pitchers, which can often be made locally, slowly leak moisture to the crop with little waste and, in effect, offer a cheap, appropriate technology alternative to drip irrigation (Gischler and Jáuregui, 1984, pp12–15). The first step with any proposed installation is a thorough check of meteorological records and conditions before construction. Small one square metre traps can be used as a cheap pilot test. The First International Conference on Fog and Fog Collection was held in Vancouver, Canada, 19–24 July 1998.

DEW HARVESTING

Dew collection as a source of water for livestock and for agriculture has attracted a good deal of study and has generated fierce disagreements about its real potential (Hubbard and Hubbard, 1907; Duvdevani, 1957; Jumkis, 1965; Dupriez and De Leener, 1992, p102; UNEP, 1983, p188). Dewponds were once common on some southern English uplands: shallow, excavated depressions, of up to 40 metres in diameter, usually sunk into the chalk bedrock. The dewpond depression is said to have been given a lining of straw, thick enough to act as an insulator to prevent heat loss at night, with a clay layer over that, giving a waterproof pond bed. On a clear night dewponds were supposed to cool more than the surrounding land, thanks to the straw insulation, while the surrounding air was warm and moist, generating dewfall. However, it seems almost certain that they mainly fill through rainfall (Hubbard and Hubbard, 1907; Dixey, 1950, p66). Ancient dew-collecting 'aerial wells' have been reported from Theodocia (present-day Crimea), the Negev, Morocco and Tunisia (Dixey, 1950, p69; Stone, 1957; UNESCO, 1962; Monteith, 1963; Balek, 1977; UNEP, 1983, p191).

However, much doubt has been expressed about the viability of trapping dew by means such as aerial wells; most are probably just piles of stones removed from farmland or runoff catchments. Certainly, those in the Negev appear to have been formed by farmers clearing slopes of loose stones to enhance runoff harvesting (Hillel, 1994, p69; Lavee et al, 1997). It is difficult to accurately measure dewfall. Pacey and Cullis (1986, p6) doubted it could provide more than a fraction of livestock or agriculture needs but that it might be of some value in establishing trees and shrubs in drylands with heavy dewfall. Evenari et al (1971) reported measuring 35 millimetres per night at one metre above ground in the Negev – a quantity that probably evaporates soon after sunrise. There has been speculation that denser vegetation might tap greater quantities, perhaps as much as 245 millimetres per night (Furon, 1963, p58).

SUBSURFACE RUNOFF HARVESTING

Sand-filled reservoirs (sand-storage dams, venetian cisterns or vegetated reservoirs) have been used in a number of countries. They can sometimes be constructed in order to collect sand over a number of years, accumulating considerable quantities of moisture-storing sediment with minimal labour input (see Figure 3.7a). Where ephemeral flows carry large amounts of sediment, this form of reservoir is a wise choice. Indeed, a normal reservoir may soon choke up. Sand-filled reservoirs can withstand flash floods better than normal reservoirs of similar capacity, can supply clean water, and are less likely to support mosquito breeding and do not suffer aquatic weed or algal growth. The storage capacity of a sand-filled reservoir is typically only 25 to 35 per cent of an equivalent sized unsilted reservoir. However, sand-filled reservoirs have the advantage that once water level falls about 60 centimetres from the surface, evaporation losses virtually cease.

The main problem in establishing sand-filled reservoirs is to ensure that only sediment of a suitable size is collected – fairly coarse sand or fine gravel. The solution is to construct a check dam of boulders, brushwood or pebble-filled gabbions which only partially obstructs streamflow, enough for coarser sediment to settle but not fine silt which would fill interstitial spaces and reduce storage capacity. These reservoirs can be used where streams or springs cannot meet short-term demand but, provided evaporation losses are reduced, can accumulate sufficiently to be useful (Barrow, 1987, pp188-190).

Percolation dams can be open reservoirs or sand-filled reservoirs. Typically, an earthen bund or check dam is used to retain overland flow or channelled runoff, so that it has time to infiltrate and recharge groundwater, where normally it would have flowed away to waste (possibly causing flood damage as it did so). Using percolation dams, springs, streamflow, and well supplies can often be improved. Other water-spreading structures such as contour seepage-furrows, *jessours*, *haffirs* and similar tank structures can also be used for groundwater recharge (see Figure 3.7b). Before construction a thorough assessment of local geology and potential pollution threats is important in order to ensure that groundwater is effectively and safely recharged. Once contaminated with sewage or harmful chemicals, groundwaters can take a very long time to recover, so it is important to prevent contamination.

Subsurface dams are useful where shallow groundwater flows run to waste. They are barriers designed to retard groundwater loss or even to improve subsurface storage. Nilsson (1988) provides a literature review and account of Swedish Agency for Research and Cooperation with Developing Countries (SAREC) and African and Indian field experience with subsurface dams. These structures have a long tradition in China (subsurface dams are quite common in the loess soil areas of Shanxi Province) and Japan (notably the southern Japanese Miyakojima islands) and around the Mediterranean (UNEP, 1983, p172). Forms developed in Sardinia and Tunisia in Roman times have attracted interest from alternative-technology development groups since the 1980s.

There are similarities between subsurface dams and methods of warping or forming sand-filled reservoirs, but a crucial difference is that, in the case of subsurface dams, the structure used to capture moisture is constructed *below* rather than above the ground surface, and storage is in a natural deposit; sand-filled reservoirs are built above ground and trap their reservoir material as it arrives in runoff (see Figure 3.7c). Some use the term groundwater dams for all structures that hold back water underground, and include within it subsurface dams and sand-filled reservoirs (WEDC, undated).

Subsurface dams can be useful for helping to recharge aquifers as well as providing a planting area. Subsurface dams can be planted with annual crops or trees, rather than storing water for livestock or crops. When water storage is the aim, the land behind the dam should be kept clear of deep-rooted vegetation which would transpire stored moisture. Subsurface dams offer the following advantages over normal reservoirs.

- There is little or no evaporation loss.
- They cannot silt up.
- There is a lower risk of water-related diseases (such as schistosomiasis and malaria).
- The land used for moisture storage may sometimes be grazed, forested or cropped.
- There is probably less risk of reservoir pollution.
- The water-retaining structure (dam), by being buried, is generally less stressed than a surface dam or bund and so takes less material and skill to build and is not so prone to catastrophic failure. Sometimes a simple waterproof membrane can be sunk in a trench and is then back filled, or a barrier of tar or cement grout is injected – both quicker and easier than building a surface dam. Nevertheless, some sediments are unsuitable and, like many of the structures discussed in this book, site selection, design and construction must be performed with care to avoid leakage or collapse, and caution must be exercised to ensure that the stored water does not cause problems to surrounding areas – such as waterlogging.

Subsurface dams are mainly used for domestic and livestock supply or limited emergency irrigation and have been constructed in many countries, such as India, Brazil, Ethiopia, Kenya, and Namibia.

Horizontal wells (or subterranean water conduit), known as the *khettara*, the *karez* or *kariz* (Pakistan), *quanat* (Iraq and Iran), *foggara* (North Africa), *falaj* (Oman and Arabian Peninsular), or whichever of its many names is applied, is a traditional means of exploiting shallow underground runoff, groundwater or springs and conveying the water, sometimes many kilometres, to a suitable site for agriculture or to a village. To construct these conduits, a 'mother well' is driven down to the groundwater source and a gently sloping tunnel is dug to connect it with suitable soil or a habitation site, often at the edge of infertile, fast-draining, piedmont alluvial-fan deposits (see Figure 3.7d). These structures are a form of underground runoff control. Their origin probably lies in what is presently Iran or Afghanistan at least 3000 years ago (Cressey, 1958; Beaumont, 1971; Wilkinson, 1974; Rahman, 1981; Sutton, 1984; Beaumont et al, 1989; Kahlown and Hamilton, 1994; Khan and Nawaz, 1995).

Today these structures are found in over 22 countries, including: Iran, India, Pakistan, Turkey, Israel (Negev), Saudi Arabia, North Africa, Majorca, Arabia, Spain, the Atlantic Islands, Cyprus, Mexico, Peru, Chile, and China. In the late 1960s they still provided 75 per cent of the water used in Iran (agricultural and domestic supplies) (Wulf, 1969) and, according to Bonine (1996), there were over 28,000 *quanat* systems in that country. Today, the breakdown of these systems is widespread and is discussed in Chapter 6. Some argue that the technique gives more chance of sustainable agriculture than is possible with modern borehole and motor-pump exploitation, although others feel it deserves to disappear because of the risks of injury to those digging and maintaining them, and because new pumps and boreholes are cheap and effective (unfortunately, effective may mean overexploitation and failure of groundwater supplies).

RUNOFF HARVESTING IN PRACTICE

Runoff harvesting has been used since the Neolithic time; for example, in Jordan it was in use by at least 9000 years BP (Bruins et al, 1986, p17). Given the history of sustained production in often difficult environments, it is amazing that traditional runoff harvesting failed to attract the attention of developers until relatively recently, and that it is still attracting little funding. Indigenous practices are still important in parts of the Middle East (notably Jordan, South Yemen and the Negev); the Americas (especially Mexico, Arizona, Colorado, Utah, Peru, and Chile); India (especially West Rajasthan); southern Russia; and Africa (especially northern Libya, Egypt, Tunisia, Algeria, and parts of the Sahel). Chapter 5 reviews the challenges faced by runoff agriculture and some of the reasons why it may break down. Runoff harvesting has often suffered because of socio-economic change: for example, in India, community self-management has been disrupted by government intervention (Agarwal and Narain, 1997). Worldwide, the introduction of tube wells and motor

(a) Sand-filled reservoir. Often constructed in annual stages to accumulate large volume of sediment for moisture storage. May also be tapped by a well. *Source:* based on Nilsson (1988), p35: Fig 5.10.

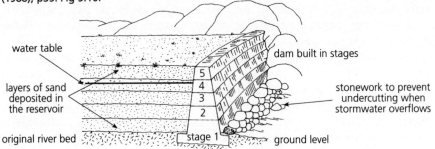

(b) *Jessour.* Essentially a dam used to capture silt and moisture so that crops (olives, dates, etc) can be grown on the moist ground. Like (a) may also be tapped with wells. Used from Tunisia to Israel. In some localities, where there are suitable aquifers; (a) and (b) may be used to recharge groundwater.

(c) Subsurface dam. The dam is constructed in a trench, or may be injected as a curtain of grout, clay, bitumen, or cement. The ground provides support; so long as there is a water-proof membrane the structure need not be very strong, and there is no risk of dangerous collapse as there may be with earthen surface dam. A sheet of plastic or other waterproof material can be used, provided foundations and capping are well-constructed. WT = water-table. *Source:* part-based on Nilsson (1990) article in *Waterlines* Vol 8, No 4 (dated 10 April, 1990).

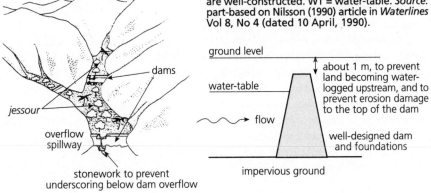

(d) *Quanat.* Although tapping groundwater, these often capture runoff from ephemeral or mountain streams that discharge into sandy alluvial fans. *Source:* sketches based on a number of sources.

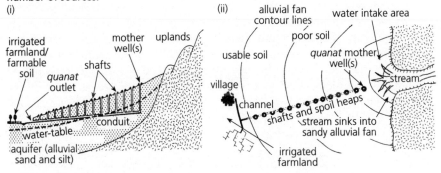

Note: schematic, not to scale

Figure 3.7 Sand-filled reservoirs and subsurface dams

pumps has proved disruptive to traditional runoff practices. Although in decline in some areas, runoff harvesting has potential, and ancient practices have provided techniques and strategies for developers to adopt with little change, or the ideas for modern 'reworked' versions (Critchley, 1989; Nessler, 1980; Van Dijk and Ahmed, 1993).

A considerable number of researchers have made comparisons of rain-fed and runoff agriculture. Rodriguez (1996) reported on studies in Balochistan (Pakistan) where the advantages of microcatchment (internal catchment-type) runoff harvesting over rain-fed cultivation depended on whether barley, rather than wheat, was grown (see similar studies by Reed et al, 1991). Furthermore, in highland Balochistan, Rodriguez et al (1996) evaluated external catchment-type runoff harvesting which supplied valley-floor planting areas. These allowed improved yields over rain-fed farming but at higher costs – which meant that rain-fed smallholder barley farmers were unlikely to adopt runoff harvesting practices unassisted. Tabor (1995) provided an overview of runoff harvesting applications in the Sahel, concluding that there had been 'spectacular' results, but that there had been less widespread adoption than was possible. For other assessments see: Singh (1985); Boers et al (1986); Carter and Miller (1991 – who reported impressive yield improvements for sorghum in poor rainfall years); Gupta (1994; 1995), and Kaushikand and Gautam (1994).

Kronen (1994) reported on progress in promoting traditional and modern forms of runoff harvesting in semi-arid areas of the Southern African Development Co-Ordination (SADCC) region (Southern Africa), as a way of improving food security for subsistence agriculturalists. The importance of a sound extension service was apparent from these studies. Runoff harvesting may only perform well enough to convert rain-fed farmers if combined with other innovations, so it is important to assess its value in combination with, say, mulching or fertilizer use (Gupta, 1989).

There are aspects of runoff harvesting for which more research would be beneficial. It would be useful to know: what are the effects of termite activity and weed growth on catchment efficiency; and what potential does runoff harvesting have for vine growing, especially where irrigation water is salty.

It is important when planning runoff harvesting to have the full participation of local people so that they can warn of potential difficulties, point out promising lines of development, ensure runoff harvesting is appropriate, and become motivated to support it fully (Cullis and Pacey, 1992). Reij et al (1988, piii) 'recommended that water harvesting be based as much as possible on indigenous techniques and local environmental knowledge'. In spite of the various problems, there has been progress in developing and promoting runoff harvesting techniques, especially in sub-Saharan Africa (Gould, 1994).

RUNOFF HARVESTING COSTS

Studies by the National Academy of Sciences (1974, p30) suggested that micro-catchments suitable for developing country smallholders could cost around US$20 per hectare of catchment. In the Negev, construction costs in the mid

1980s were between US$10 and US$40 per hectare cultivated (Postel, 1985, p4). For Kenya, Pacey and Cullis (1986, p156) suggest microcatchments require 600 man hours per hectare cultivated to construct *and* 320 man hours per year per hectare cultivated to maintain. In that part of the world traditional rain-fed cultivation typically requires about 200 man hours per hectare cultivated per year. To be weighed against these figures is the improvement of yields from as little as 45 kilogrammes per hectare to as much as 2000 kilogrammes per hectare cultivated (Lewis, 1984).

Reij et al (1988, pp69–74) give an indication of labour inputs for construction and maintenance and some idea of costs for various forms of runoff harvesting in Ethiopia, Kenya, Niger and Burkina Faso. Pandy (1991) examined the economics of runoff harvesting in semiarid India. It is clear from a number of studies that caution is needed. Firstly, it is often unclear exactly how estimates of labour input and quantities of material are arrived at, and units may not be standardized ('loads' of stones or brushwood are vague and, as discussed in Chapter 3, man hours or person hours are often imprecise measures). Secondly, the advantages of runoff harvesting over rain-fed agriculture may not be immediately apparent or necessarily attractive to potential practitioners.

4   EPHEMERAL AND VARIABLE STREAMFLOW: SPATE, WETLANDS AND FLOOD AGRICULTURE

This chapter deals with floodwater farming, *wadi* agriculture, flood spreading, spate irrigation, flood recession agriculture and storage of floodwater (note: flood irrigation generally refers to water application by the flooding of fields – supplied with canal water or groundwater but not floods). Spate, wetlands and flood agriculture overlap with runoff harvesting methods discussed in the last chapter. However, the latter mainly seek to utilize flows that have yet to reach a stream; this chapter deals with the use of runoff which has generally reached a stream or river, or which has flooded an area. Spate and flood agriculture focuses on coping with periodic, often sudden, heavy debris-charged flows; the approaches discussed in the last chapter focus more on water spreading – holding relatively easy-to-control flows where they can infil-trate or be stored (FAO, 1987).

A crude division can be made into:

- *spate agriculture* (see also *wadi* agriculture in Chapter 3) – using flows from ephemeral streams and often seeking to control stream erosion as well as to obtain water;
- *wetlands and swamp agriculture* – using land that periodically drains enough for agriculture or which can be modified to support it (some of these are relatively small wetlands; others, such as the *Sudd* of the Sudan and Egypt or the inland delta of the River Niger are vast);
- *river flood agriculture* – using the floodwater of larger rivers; in some regions these floods are quite regular and predictable enough for cropping strategies to be planned around them – for instance, cropping the *várzeas* of eastern Amazonia, *warping* (see Chapter 3), and water-meadows (Cowan, 1982);
- floodwaters may also be intercepted and spread to improve groundwater recharge.

Spate and flood flows are often important for agriculture in regions where there is inadequate rainfall – a huge diversity of strategies and techniques have evolved in Africa, the Middle East, Afghanistan, Pakistan, Nepal, Bhutan, Egypt, Algeria, Morocco (where Mabry, 1996, p6, estimated seasonal flows and floods support two-thirds of all irrigated land), the Andes, and many other regions (Lawrence, 1986; Van Immerzeel and Osterbaan, 1990; Morton and Van Hoeflahen, 1994). Spates and floods may result from periodic, intense rainstorms, extended (usually seasonal) rainfall or from snow melt or glacier meltwater. These sources of water may be distant from the areas receiving the floodwater, often in mountain areas.

There is a rich tradition of these forms of agriculture, some of them of very ancient origin. Many of these indigenous strategies have the potential for wider use and might possibly be improved upon and spread (Bryan, 1929; Kahlown and Hamilton, 1996). Larger spate and flood irrigation structures, such as the bigger tanks and bund systems, demand the cooperation and coordination of a number of people for construction, regular maintenance and repair (Van Steenbergen, 1997). Some of these strategies are amongst the world's most productive forms of agriculture, include the longest-sustained productive farming strategies, and in most cases require little input, other than floodwater and human labour, to maintain fertility.

Caution is needed in exploiting floodlands and wetlands because these are often rich in biodiversity and can be easily disturbed, even by limited agricultural development. Also, like rain-fed agriculture, use of floods and seasonal runoff (especially where the latter is not linked to storage tanks or cisterns) often involves risk of failed harvest and even land damage. Care and sound practices reduce these risks for runoff agriculture, but for rain-fed agriculture do little to counter 'uncertain rain'. Wetlands and floodlands comprise at least 6 per cent of the world's land surface and have come under considerable pressure from development. Wetland exploitation needs to be tempered with more concern for conservation, and efforts should be made to adopt management strategies that allow for biodiversity maintenance – for example, provision of feeding and resting habitats for migrant birds (Roggeri, 1995, p38). The Convention on Wetlands of International Importance, more commonly known as the Ramsar Convention of 1971, is a global treaty providing the framework for protection of wetland habitats which are important for migratory fauna and for the well-being of traditional land users (IUCN, 1982).

SPATE AGRICULTURE

Spate agriculture (*wadi* agriculture, or floodwater farming) uses floodwater flowing in ephemeral or highly variable streams which can be utilized to irrigate plots to support crops, pasture or trees (Stern, 1979, pp51–55; Gilbertson, 1986). There is a tradition of using flash floods in Asia, South Asia, the Middle East, parts of Africa, and the Americas (Brunner et al, 1986). Some term this *wadi* agriculture (in the Americas it is known as *arroyo* agriculture).

A degree of risk is involved because streams may flow forcefully and destroy flood collection or spreading structures and the water may be heavily charged with debris which can choke channels and bury cropland. Stream courses may shift from flood to flood, and the timing and severity of flood events may be difficult to predict. Also, floodwater originating from melting snow or glaciers may be cold enough to damage crops and is sometimes charged with harmful sediments (Butz, 1989).

Impressive spate-farming systems were developed by the Papago and the Navajo peoples of the Sonoran Desert, Arizona, southern US and Mexico, and by other peoples in Latin America, Africa and Asia. These systems generally spread floodwater and silt from seasonal *arroyos* (*wadi*-type channels), or used overland flow from bare hillsides. More recently others have exploited the use of runoff from highways. Although some of this spate agriculture has been abandoned, there are still a sizeable number of cultivators practising it around the world (Nabhan, 1986a; 1986b; Doolittle, 1989).

Check dams have been used for centuries in India, China, Sri Lanka, and Latin America. Check dams (retardance dams or gully plugs) are a means of trapping and using silt-charged floodwater flowing in channels, or of diverting flows for agricultural use. They can also be used to counter gully erosion and rehabilitate gullies. Cheap to construct, check dams are useful in marginal conditions where farmers have little money (Dennell, 1982). They can extend the duration of streamflow (downstream) and can also be a way of using flood-water to recharge groundwater (see percolation dams in Chapter 3). They can be used in steeply sloping channels, to control gully erosion and improve infil-tration, and may capture soil to provide moist planting areas (see *jessour* in Chapter 3). These qualities can be of value to conservation services who wish to improve habitats for wildlife (Debano and Schmidt, 1990).

Spate agriculture provides a livelihood for a large number of relatively poor people in marginal environments; often (as, for example, in parts of Pakistan) it is the predominant cropping strategy (Van Steenbergen, 1997). There is a diversity of construction techniques, all designed to slow flows and trap debris if possible without suffering catastrophic damage during severe floods (see Figure 4.1). Unfortunately, they do not always escape failure, especially if not well designed, carefully sited and soundly constructed. For example, Santiago Island in the Cape Verde islands had about half of its check dams fail in storms during 1984 (Moldenhauer and Hudson, 1988, p27).

Broadly, there are two types: impervious construction with strong spill-ways to dispose of peak flows; and semipermeable construction that leaks enough to survive floods and which can be relatively easily repaired if damaged. The latter type tends to be constructed from locally available materi-als such as stones, stone-filled gabbions, brushwood, logs, old sacks filled with sand, coarse-wattle hurdles or old car tyres (Tuan, 1988). While check dams can be very useful for conserving silt and moisture, they have disadvantages:

- they do not necessarily cure causes of erosion (they treat symptoms);
- the moisture they store may not be enough for a crop to mature for harvest if floods are infrequent;
- there may be water-related disease problems associated with ponding (less likely with semipermeable check dams) (Goudie, 1990, pp254–257).

(a) plan

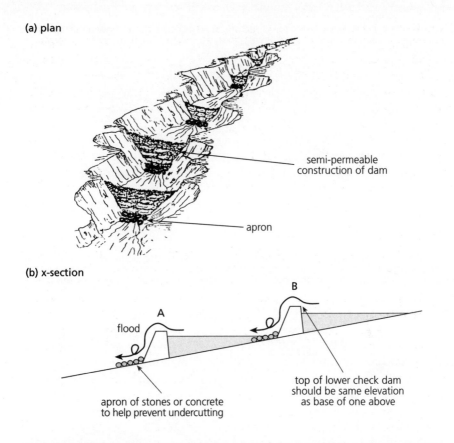

semi-permeable
construction of dam

apron

(b) x-section

B

flood

A

apron of stones or concrete
to help prevent undercutting

top of lower check dam
should be same elevation
as base of one above

Note: height of check dams, see (b) depends on construction materials (sand-filled sacks; stone-filled *gabbions*; brushwood; stakes and wattles; stacked sods; logs). Also, the top of A must not be much below the base of B, or B will be at risk of being undercut, leading to collapse and progressive collapse of upvalley dams as a consequence. Construction of the check dams should be semipermeable; a protective apron of stones or concrete at the base of each dam helps prevent undercutting during floods. Positioning of check dams is thus as important as quality of construction.

Figure 4.1 Check dams

There can be vegetative alternatives to check dams; barriers of robust plants such as the agave species can control gullying and act as check dams to accumulate soil upslope (Rajeev, 1992). A barrier of freshly cut faschines (branches which are interwoven to form a barrier across a channel) will often set roots and grow to form a substantial and well-anchored barrier, ideal for erosion control or stream diversion, but less useful where the goal is to accumulate a planting plot of wet soil (because the roots take much of the moisture).

Water-spreading structures are valuable where there are sudden, silt-laden streamflows and it is difficult to build headworks (offtake structures which divert streamflow to a channel or pipeline) that will not get damaged or choked up. One solution is to end a channel sufficiently away from the streamcourse so

(a) Check dam structures in *wadi* to spread and slow periodic streamflow and trap debris. Once silted-up can be planted with olives, dates, etc (for similar structures see: *jessour*).

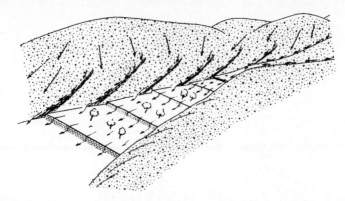

(b) Diversion from a *wadi*. Detail of the vital headworks, in this case a *khul* – designed to survive damage and feed water without silting up. *Khuls* are inundation canals (sometimes called *pynes*, at least in Sind province); these are common, especially along the Indus River.

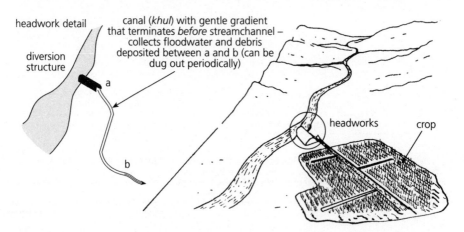

(c) Diversion structure in *wadi* and spreader-storage bunds. Diverted water led onto floodplain. In Somalia these systems are known as *caag*, in Yemen as *say*, in Tunisia as *jessour* (or *gessour*). In the Americas the *arroyo* systems are similar.

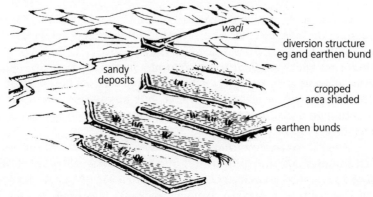

Note: schematic, not to scale

Figure 4.2 Using spate and *wadi* flows

that it fills in time of flood with water that has risen above the stream but has lost much of its energy and debris load. Such structures are common in Kashmir, western Bengal, (kuhls), and north-west India and Pakistan (*pynes*) (especially in the Sind). The channel is slightly sloping along the contour (typically with a gradient of 1 in 100) to a suitable cropping or pasturage area (see Figure 4.2).

STORAGE OF FLOODWATER AND HEAVY RUNOFF

Most approaches to storing occasional runoff flows rely upon earthen bunds, which means smallholders do not have to acquire costly construction materials.

Earth dams and *tank storage* can take a range of different forms. *Khadins* and similar *liman*-type earthen-bund infiltration basins and dams have been discussed in Chapter 3. Similar structures can be used to spread or store flood-water from channels (Vlaar, 1992). Earthen bunds that were used to exploit river floods include the *horse-shoe dike* (see Figure 4.3a). Even more simple is the use of natural depressions, cropped as floodwater recedes (the crop planted to 'pursue' the falling watertable). An example of this approach is the *dhasheeg* agriculture practised in Somalia and parts of the Sudan. This provides secure, high and sustainable yields in a region where rainfall is variable and unreliable. Demanding few modern inputs, the approach should be possible wherever there are topographic depressions with suitable soils that can be flooded from a river or by slopewash (Besteman, 1996) (see Figure 4.3b). With modern earthmoving equipment similar structures could be excavated.

In south Asia (north-west India, south India, east India and Sri Lanka) the term tank is widely used; elsewhere earth dams, small reservoir or on-farm reservoir are broadly synonymous. *Tanks* are shallow reservoirs usually impounded behind simple earthen bunds, with a weir or drop structure to shed excess water, hopefully avoiding overtopping and catastrophic break-down of the impoundment structure. Tanks are sited where they can intercept flows from seasonal channels, or by the side of a river or *wadi* to catch flood-water. The risk of flood damage is usually reduced by ensuring a carefully designed and positioned channel which also reduces silt content. Typically, a tank feeds excess water, once full, to another downslope, and so on. Tanks are nowadays mainly found in Sri Lanka and India and are a good way to exploit heavy seasonal (monsoon) rainfall which would otherwise flow to waste. The tank may provide dry season irrigation, livestock watering, domestic supply and perhaps some fish. Building costs are quite low and they can be constructed relatively rapidly; they also provide domestic water and can support pisciculture.

In 1963 tanks provided more than 10 per cent of the irrigation needs in most Indian states and in some over 40 per cent, with an average of nine hectares irrigated for each tank (Goudie, 1990, p258). Overall, perhaps 40 per cent of east Indian irrigation is by means of tank systems. Roughly 3.3 million hectares of land were irrigated by tank in India in 1988. In Sri Lanka in the mid

(a) Horse-shoe dike. Earthen bund used in area that floods. Captures silt (and helps reduce flood damage downriver by spreading floodwater). Common on floodplain of Irrawady (Myanmar) and in delta lands of Bangladesh. After floodwaters fall depression is cropped for rice (some are over 250,000 ha in size). *Source:* based on Roggeri (1995), pp204–205, Figure B2.

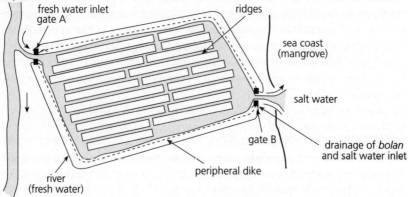

(b) *Bolan.* These are paddy fields reclaimed from mangrove swamps. Widespread in low-lying coastal areas of West Africa. The technique seeks to reduce salt content of the ridges – in the early rainy season salts are leached out of the ridges (salt tends to accumulate there in the dry season due to capillary rise of seawater which seeps in); at low tide gates (A and B) are opened, letting saltwater escape. *Source:* based on Dupriez and De Leener (1992), p269.

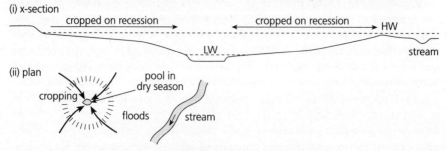

(c) *Dhasheeg* agriculture. An indigenous flood recession agriculture system used in Somalia. Similar approaches are found worldwide. The shallow depressions flood in the wet season from streams and are cropped as groundwater falls in drier periods. A pool may remain which is often used for dry season irrigation. Valuable on heavy clay soils which tend to shed water without infiltration (Besteman, 1996 written description – no illustrations).

(i) x-section

(ii) plan

Figure 4.3 Flood recession agriculture

1970s there were 132,000 hectares (both estimates are probably too conserva-
tive) (Ambler, 1994, p270). A number of researchers have looked at the
potential for introducing tanks to areas beyond south Asia where conditions
seem suitable and there should be benefit from water storage for livestock or
crops (Intermediate Technology Development Group, 1969).

The main disadvantages of tanks are:

- there is unpredictable variation in the quantity of water they store each
  year;
- they take up valuable space in valley bottoms where some of the best
  farmland lies;
- they may provide breeding sites for disease-carrying insects and other
  pests;
- their banks and spillways need to be constantly maintained to prevent
  leakage and slumping, and regular cleaning out of silt and aquatic weeds;
- there is often considerable seepage and evaporation loss from tanks, and
  water quality usually deteriorates as the year progresses.

Tanks are easier to clean of silt if allowed to empty at the end of the dry season
so that the silt can be dug out and carted away for use on farmland as compost.
Unfortunately, in south Asia, with the spread of chemical fertilizers and more
employment opportunities for rural labourers, there is less incentive for this
sort of tank cleaning. Perhaps if agrochemical-related pollution becomes a
problem there may be revived interest in using tank-silt on farmland. SWC
measures in a catchment can support tanks, helping to regulate the rate of
runoff and reducing the silt content of flows. This is particularly important
where catchments that once yielded clear flows now rapidly silt tanks; often
this is because of damaging shifting cultivation, poorly managed forestry or
other forms of land degradation (Dharmasena, 1994). A promising development (pioneered by ICRISAT, India) is the raised bed and grassed-channel
system feeding tank(s) with less silty runoff than traditional layouts. The tank-
stored water can be used to irrigate the raised beds using watering cans, pump
and hose or bullock-drawn sprinkler cart.

South Asian tanks are usually communally owned, which means building
maintenance and water allocation must be organized by a group of people
who are often slow to respond to problems. During India and Sri Lanka's
colonial era, the British did little to sustain tanks or to stimulate interest
(Ambler, 1994). Today the situation is largely one of neglect. In parts of South
Asia the controls exercised by traditional rulers and communities have relaxed,
especially since the 1960s; public works departments are often responsible for
tank maintenance and use, and as farmer involvement has decreased, the
quality of management has usually fallen (Rao and Chandrakant, 1984; Von
Oppen and Subba Rao, 1987; Stern, 1988; Palanisami and Easter, 1987;
Govindasamy 'and Balasubramanian, 1990; Palanisami, 1990; Shankari, 1991).
The breakdown of tank storage, and the upgrading of tank agriculture, are
discussed in Chapter 6. The amount of water stored by a typical tank varies
widely from year to year, as rainfall fluctuates, and this has prompted attempts

to model and predict water availability, to improve design, and to assess future prospects (Mahendrarajah et al, 1996; Govindrasamy, 1997; Govindrasamy and Balasubramanian, 1997; Verma and Sarma, 1990).

WETLANDS, SWAMP AND FLOOD AGRICULTURE

Richards (1985, p29) described flood irrigation as natural irrigation, as opposed to technocratic, artificial, mainstream irrigation; the same could be said of wetland and swamp agriculture. Although widespread, often with a tradition dating back thousands of years and in many localities with great potential, wetland, swamp and flood agriculture has attracted less study and investment than mainstream, canal-supplied gravity irrigation, groundwater-supplied irrigation or pump irrigation (Beseman, 1996, p53).

Wetland and seasonal swamp agriculture is already important in some regions and offers considerable potential for expansion. Suitable areas exist, even in the driest environments, and these can be very productive. These areas also offer opportunities for sustaining crop production, even if there is little input of agrochemicals or manure, thanks to inflow of detritus and dissolved nutrients and the nitrogen-fixing activity of aquatic bacteria and algae. Some ancient civilizations have been primarily founded on wetland development, or at least fed by it – for instance, the Maya of pre-Conquest Latin America and 'hydraulic civilizations' along the Tigris, Euphrates and Nile rivers. Some wetlands are vast – for example, the inland delta of the Niger varies in extent between 4000 and 30,000 square kilometres; the Kafue flats (of Zambia) expand to 28,000 square kilometres in the wet season. Other large wetlands include the Pantanal (South America); Lake Chad region (West Africa), the *várzeas* of Amazonia (Anderson and Jardim, 1989; Roggeri, 1995, p3), the Sudd swamplands (Sudan and Egypt), and the coastal wetlands of Guyana.

There are many types of wetlands, some easily exploited for agriculture, others presenting challenges because of their difficult soils, unpredictable flooding, or problems with tree-stump removal. For a recent review of tropical freshwater wetlands, see Roggeri (1995). For a general introduction to the world's wetlands and their management see Mitsch and Gosselink (1993); for case studies of sustainable development of wetlands and wetland management (in the UK, France, Spain and the US), see Turner and Jones (1991).

Large reservoir *drawdown* areas are man-made wetlands, created when reservoirs reach their lowest capacity before refilling. These can sometimes support productive agriculture and offer a means of providing those relocated from the inundated areas with alternative cultivable land (Chaudhry, 1982). Lake Volta (Ghana) alone is reputed to offer over 90,000 hectares of drawdown area, which so far have been little developed (Kalitsi, 1973). In the past reservoir management authorities have tended to discourage use of drawdown areas, mainly for fear of reservoir contamination and risk of transmission of water-related diseases. However, there have been informal and officially supported attempts to develop drawdown cultivation with reasonable success. There are some cases where the reservoir management regime means that

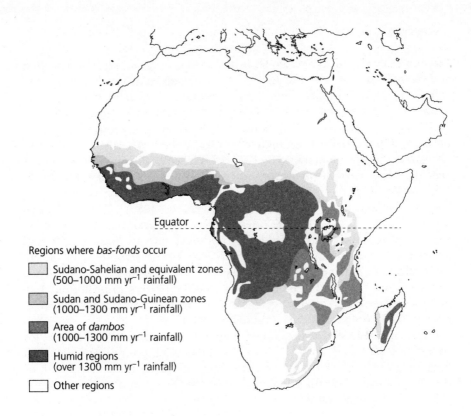

Regions where *bas-fonds* occur

- ☐ Sudano-Sahelian and equivalent zones (500–1000 mm yr^{-1} rainfall)
- ☐ Sudan and Sudano-Guinean zones (1000–1300 mm yr^{-1} rainfall)
- ☐ Area of *dambos* (1000–1300 mm yr^{-1} rainfall)
- ☐ Humid regions (over 1300 mm yr^{-1} rainfall)
- ☐ Other regions

Source: Adams and Carter (1987), p5: Fig 1

Figure 4.4 Regions in tropical Africa where *bas-fonds* are found

refilling may take place at unpredictable times, making agriculture too much of a gamble (Adams, 1985a).

Bas-fonds are shallow, seasonal valley swamps, found throughout sub-Saharan Africa. They are formed where there are depressions and the watertable lies at or near the surface for part of the year (Raunet, 1985a; 1985b; Denny, 1993) (see Figure 4.4). In wetter regions high watertables are responsible for the flooding; in drier it is usually storm runoff. Generally, there is no clear drainage channel and slow flow takes place through dense grass, sedge and herbs. These wetlands can be several square kilometres in extent and are often exploited for household garden cultivation or rice cropping (Raunet, 1984). As populations have increased *bas-fonds* agriculture has become increasingly important and in dryland areas provides a valuable way of producing subsistence crops. Some have expressed the hope that more development of these areas might help counter rural–urban migration (Zanen, 1995). A number of other types of African wetlands (discussed in the following section) are seen by Raunet (1985b) as subtypes of *bas-fonds* and also support considerable agricultural production.

Dambos

Dambos (a Bantu word) are also known as *vleis* (Africaans), *mabuga* (Swahili) and *matoro* (Shona); they are sub-Saharan wetlands traditionally exploited in some areas, but in others unused and offering considerable potential (see Adams and Carter, 1987, pp6–7). *Dambos* (the term mainly used today) are shallow, grassy swamps in valley bottoms which drain gently toward rivers and naturally collect runoff from higher ground in the wet season as runoff or subsurface seepage. Sometimes they have fertile, cultivable soils or sometimes clay soils which are difficult to work and drain. *Dambos* occur throughout sub-Saharan Africa where there is more 1000 to 1300 millimetres rainfall per year and the soils and topography are suitable (Richards, 1985, p72; Owen, 1995; Reij et al, 1996, p114). They are often important for grazing, domestic water supply and household garden cultivation, although they are likely to be part of an overall livelihood strategy which may also include rain-fed cultivation, pastoralism, and sometimes shifting cultivation (Bell and Roberts, 1991). Hotchkiss and Lambert (1987) estimated their total extent to be around 1.25 million hectares; they are especially common in Zimbabwe (where approximately 200,000 hectares were cultivated in 1987).

The soils of these wetlands do not drain well naturally, and so, at least for part of the year, they are waterlogged (Stern, 1979, p34). If drained, or if farmers can plant a *dambo* as the watertable falls, crops can be sustained, or the *dambos* can be used for fodder production. There are three main strategies of *dambos* exploitation:

- wait for natural drainage to recede, allowing 'flood recession agriculture';
- dig ditches to speed drainage;
- excavate suitable pits or channels from which water can be scooped up to irrigate raised beds (similar to the South American *chinampas* systems) (Lambert, 1990).

Fadamas

Fadama refers to land in Hausa-speaking areas of Nigeria which are seasonally waterlogged or flooded (Adams and Carter, 1987, p7). These may be small depressions, features similar to *dambos*, or the floodplains of large rivers; however, they differ in that they are not perennially wet swamps. Exploitation strategies are similar to those used for *dambos*.

Bolis

Bolis are *dambo*-like wetlands or backswamps of river floodplains. Some have enough water to support more than one harvest of rice or other crops per year.

WET FIELD, DRAINED FIELD AND RAISED FIELD OR RAISED BED AGRICULTURE

Chinampas (*waru waru* – Peru; *tablones* – Mexico) are the best known of a number of very effective wet field and drained field agricultural systems, developed in pre-Conquest Central and South America. In Rwanda an equivalent strategy is called *hotillonage*, and similar traditions are to be found in Bengal, Belize, Bolivia (around Lake Titicaca), Zimbabwe, and the lower Tigris–Euphrates basin (Iraq). *Chinampas* consist of raised beds, or in some areas floating gardens of soil built on reed rafts, separated by channels (see Figure 4.5a) (for a bibliography, see Roggeri, 1995, p208; Ericksson, 1989). The raised beds are formed by excavating material from the channels onto planting areas with the addition of layers of vegetation (such as reeds and aquatic weeds) to raise the planting site above flood-level and waterlogged soil. Periodically the cultivators clean the silt from the channels, wading or working from a boat or canoe, and add it to the bed with fresh layers of vegetation. This compensates for subsidence and shrinkage, caused by decay and compaction, and renews fertility.

Chinampas produces a very diverse range of crops from the raised beds, as well as fish, ducks and aquatic plants from the channels – in effect, it can be a very productive integration of agriculture and aquaculture. The system is also highly sustainable, even in difficult environments, and makes use of difficult-to-drain wetlands (and can be used along frequently flooded lake margins with the planting plots built on reed rafts). It ensures that the crops can get moisture by rooting down into the lower, wetter part of the bed, and offers protection against night frosts which can be a problem in Andean South America (Wilken, 1987, p43). This adaptability to difficult-to-drain soils and protection against cold and flooding should mean that the approach has potential for a diversity of areas in Asia, the Americas, Africa and Europe. *Chinampas* agriculture produces a considerable diversity of crops – perhaps 30 species in a locality (Armillas, 1971; Darch, 1988; Crews and Gliessman, 1991; Jiménez-Osornio and Gomez-Pompa, 1991; Sluyter, 1994).

The Maya were constructing *chinampas* roughly 1000 years ago, and the system flourished, supporting quite large populations, between about AD 100 and 700 (Skarie and Bloom, 1982). The Aztecs used similar techniques, cropping swamps in what is now central Mexico, especially the Veracruz region; around Lake Texcoco alone there was an estimated 10,000 hectares which fed over 100,000 people for a long period of time. Surviving *chinampas* agriculture (for example, near Mexico City) confirms that the approach offers great potential for sustainability with little outside input, and can give good crops, perhaps better than modern alternatives. There is potential for expansion and an advantage in that the *chinampas* system offers an alternative to land drainage and conversion to 'normal' agriculture, which may cause off-site impacts and can sometimes go horribly wrong, leading to soil degradation. However, some areas are unsuitable if the peats or soils generate methane or suffer acid–sulphide problems.

(a) Raised bed system (*chinampas*). Can sustain agriculture in waterlogged areas, even extending into swamps and lakes. These lands may be rather un-productive before conversion. The approach may also counter frosts and other unfavourable climatic conditions. Very productive, diversified cropping, and sustainable without inputs from outside locality. Can be integrated with aquaculture

(i) x-section

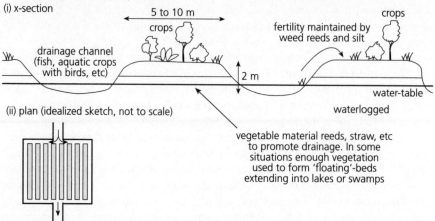

(b) Agro-piscicultural system. Similar to (a), but with larger ponds. As with (a) material cleaned from ponds/channels used to build up and maintain fertility of raised beds. Requires sufficient drop of slope and an inflowing stream to ensure throughflow; (a) more likely to be level swamp, peatland or lake margin. *Source:* redrawn with modification from Dupriez and De Leener (1992), p76: Fig 119

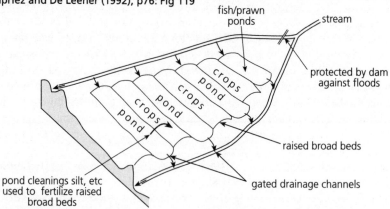

(c) Flood recession agriculture. River levée (naturally formed and periodically submerged by flood) is cropped at progressively lower elevation as floodwater falls. Often modification with bunds and flap gates to control recession and form rice paddy. In some estuaries tides may back up water and allow limited and predictable flooding at frequent intervals during low riverflow (LW), so irrigation water can be renewed. Annual flood may deposit fertile silt and destroy weeds and pests, making approach sustainable with little outside input. *Source:* based on author's field research in Amazonian *várzeas*.)

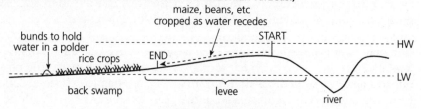

Figure 4.5 Wet field techniques

There are a number of strategies similar to *chinampas* – for example, agropisciculture, using broad beds and ponds fed from a valley stream, can be a productive and sustainable strategy for crop and fish or prawn production. Floods can be diverted down the lower channel to avoid damage and the throughflow of water helps to reduce the problem of downstream pollution caused by waste from aquaculture (see Figure 4.5b).

RIVER FLOOD AGRICULTURE

Roggeri (1995) divided tropical freshwater floodlands into: river floodplains; tidal floodlands (where sea tides back up freshwater); deltas; mangrove swamps; and valley bottom floodlands. Richards (1985, p78) suggested the division of flood agriculture into 'flood advance' (rising) and 'flood retreat' (recession) agriculture, a division Dupriez and De Leener (1992) also favour. *Flood recession* agriculture predominates, probably because agriculturalists are more certain of the behaviour of falling water levels. *Flood recession* agriculture (floodplain farming or *décrue* agriculture) involves reliance of moisture left in the soil as floodwaters recede (see Figure 4.5c). Many rivers have large enough, sufficiently predictable and favourably timed flooding to support flood agriculture. The Niger River alone has over 6000 square kilometres of flood-adapted agriculture (Adams, 1992, p19). There are traditions of flood recession cropping or grazing in many regions, notably: the inland delta of the Niger; the *Sudd* swamps (which cover over 155,000 square kilometres of the Sudan); the floodlands of the Omo Valley (Ethiopia), the Tana River (Kenya), the Rufiji River (Tanzania), and the Zambezi and Lufira rivers (Zaire), as well as extensive areas of Egypt; the coastal lands of Guyana and Bangladesh (Roggeri, 1989); and the extensive *várzeas* of Amazonia.

Flood advance agriculture is mainly restricted to rice cultivation using flood-resistant varieties developed in south Asia, South East Asia and West Africa (see later in this chapter), and cultivation of floodlands where flooding is very predictable (such as some of the Amazonian *várzeas*). Flood agriculture can involve quite a high risk of crop loss if floods come early or farmers have delayed harvesting.

Some peoples have closely adapted their livelihood strategies to flood regimes – for example, in parts of Amazonia, the marshlands of Iraq, in parts of Bangladesh (Paul, 1984), in the *Sudd* wetlands of Egypt and the Sudan, and along the Nile, Benue, Niger and many other rivers. Where flood recession agriculture is traditional there has often been recent disruption, caused by the construction of large dams or barrages which have altered natural flooding. Sometimes this control of natural flooding has been so drastic that agriculture, fisheries and wildlife have suffered badly. This has prompted researchers and a few development authorities to experiment with artificial floods (freshets): controlled releases from dams (Roggeri, 1995, p60). Controlled releases may be difficult or impossible to organize where the authorities are maximizing electricity generation or dry-season water supplies. When controlled releases are possible, flood agriculturalists and wildlife conservation bodies should cooperate to maximize benefits.

(C J Barrow, 1985)

Photograph 4.1 Flood recession agriculture in the *várzeas altas* of the
Tocantins River, Amazonia. Smallholder plot during
low-water season (from low water to maximum flood level is about 12
metres, as can be judged from small boat on right), roughly
20 kilometres below the Tucuruí Dam, Amazonian Brazil

Another aspect of floodplain and wetlands development is that cropping may
compete with established pastoral, fisheries and other usage, and with aquatic
and river-margin wildlife. This conflict may need to be carefully monitored and
controlled, especially the impacts of agrochemical use since this may easily
contaminate wetlands and rivers, causing serious damage. In the *Sudd*
wetlands there is a tradition of seasonal pastoral transhumance; cattle herders
move onto pastures that flourish after the water recedes, and conversion to
arable farming would probably come into conflict with those seeking grazing.
In Amazonia there is a similar pastoral transhumance tradition, with herders
who use *várzeas* pastures moving their livestock in the flood season to dryland
or stalls raised on piles (*marombas*) until they return to graze floodland
pastures as waters fall; flood-recession irrigation development may therefore
come into conflict with herders.

By constructing earthen bunds with simple inlets and valves, it is possible
to better control flooding. *Polders* can be formed which trap water as a river
floods, but prevent excess flooding once inlets are shut; the water may then be
retained after the river flood subsides (see Figure 4.6). In some wetlands where
flooding is tidal (backing up freshwater in lower courses of rivers), bunding
and valves can allow quite precise control of water level and a regular renewal
of water (this sort of *polder* is used in eastern Amazonia's *várzeas*) (Lima,
1956).

(C J Barrow, 1998)

Photograph 4.2 Flood recession and water spreading agriculture during low-flow season in the High Atlas Mountains, Morocco

FLOOD-TOLERANT RICE AND FLOATING RICE

Rice is grown in wetlands and floodlands as well as in carefully regulated, shallow-water rice paddies in Asia, South East Asia and West Africa. There is a rich indigenous knowledge of floodplain farming in west Africa, especially along the Niger, Sokoto, Rima and Senegal rivers (and many others) (Grove, 1985, p170). West African deepwater rice (*Oryza glaberrima*) has been cultivated for as much as 2000 years and there are similar forms of Asian rice (*O. sativa*), able to cope with slowly rising floodwater. Some rice varieties will stand brief submergence without much damage (*flood-tolerant* rice). *Floating rice* varieties grow fast enough and sufficiently long (stems can reach 2–3 metres in length) to keep up with rising floodwater (Richards, 1985, pp75–77).

AQUATIC CROPS

Where runoff forms lagoons or marshlands there may be potential for harvesting aquatic plants such as water hyacinth (*Eichornia* species) or other aquatic weeds and algae for compost, charcoal or alcohol production (Roggeri, 1995, p31). Emergent species such as reeds (for example, *Phragmites* species) can also provide useful supplies of construction material, or support paper and fibre-board making. Runoff, especially that contaminated with livestock manure, human sewage or other organic waste, is likely to be too rich in nutrients to safely discharge directly into streams or other water bodies without the risk of

(a) Simple flood recession agriculture. The finer silt is deposited away from main channel where flow is slower. In some floodplains farmers try to slow floodwater to deposit fine silt (marling). Not all rivers deposit suitable marl deposits. *Source:* based on author's field research in Amazonia.

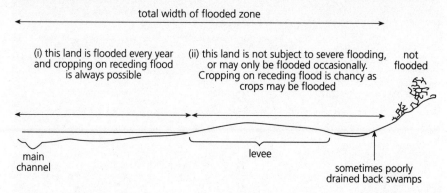

(b) Modification of floodlands to form *polder*. Simple earth bund with flap-gates (and sometimes a pump to raise water); can be used to create rice paddies or aquaculture ponds

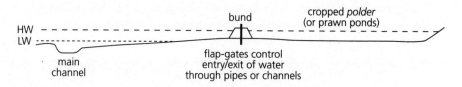

Figure 4.6 Flood recession agriculture

excessive biological oxygen demand (BOD) and other problems. If such contaminated runoff is fed to a site where it can flow gently through aquatic vegetation, microorganisms will help to reduce the nutrient content. A treatment lagoon or marsh can yield reeds or aquatic weeds with the potential values already mentioned, and will also largely purify the runoff. There are parts of the world where irrigation return flows or other forms of runoff are contaminated with salts, boron, arsenic, or various heavy metals (for example, in parts of the Grand Valley, California). If these are treated in lagoons with tolerant reeds or other plants, and the biomass can be harvested and safely disposed of, wildlife and water bodies will be protected from pollution.

5 INDIGENOUS RUNOFF AGRICULTURE: CHALLENGES AND BREAKDOWN

Agenda 21 (Chapter 26) called for:

> *...involvement of indigenous people and their communities at the national and local levels in resource management and conservation strategies and other relevant programmes established to support and review sustainable development strategies (for an introduction to* Agenda 21, *see Keating, 1993)*

In recent years there has been growing interest in indigenous knowledge systems and it has begun to make some impact on established development approaches, although these still tend to ignore community-based knowledge and local habitat needs and potential. Attitudes might change as a consequence of the Declaration of the Indigenous People's Earth Charter, signed on 30 May 1992 (IUCN, 1997, p179), but that change is likely to take time. Indigenous knowledge is especially useful for remote areas where it is unwise to depend on outside inputs, and where aid in times of scarcity is likely to be unforthcoming or slow to arrive (Agarwal and Narain, 1997a; 1997b).

Indigenous runoff agriculture has a long history; some societies developed strategies which successfully sustained crops or pasture for centuries, even in harsh environments. However, indigenous runoff agriculture has been overlooked by planners and underresearched by those seeking to improve agriculture; indeed, before the late 1960s it was virtually ignored by development agencies (IFAD, 1992, pp84–5). Leach and Mearns (1996) comment on the way colonial powers overlooked indigenous SWC and supported 'modern' SWC, often of an inappropriate kind that was promoted in a clumsy manner, with the result that development efforts have frequently been environmentally damaging and unsustainable.

Worldwide indigenous runoff-agriculture strategies, some of which have much to recommend them, have in the last few decades broken down or are today in the process of degenerating (Pangare, 1992). A number of countries, once self sufficient or even exporters of food, were importing cereals by the early 1990s and showed signs of further decline. For example, Morocco, Tunisia and Algeria exported grain until roughly 20 years ago, but by 1992 imported 20 per cent, 40 per cent and 60 per cent (respectively) of their cereals. A growing number of other countries, especially in sub-Saharan Africa, have undergone a similar shift. Indigenous food production must be protected and improved.

WHY DOES BREAKDOWN OCCUR?

It is not only runoff agriculture that is facing challenges and suffering breakdown; other established livelihood strategies are in difficulty or have failed. The causes are numerous, often complex and indirect, but include social change; the encounter with the modern bureaucratic state; the spread of capitalism; globalization; population increase; loss of access to common resources; civil unrest; the introduction of new technology (such as the plough – making hand tool-maintained runoff farming less attractive), and structural adjustment programmes (see Box 5.1). Pereira (1989, p3) summed up the situation as 'subsistence agriculture under stress'. Throughout the world until recently the family unit has been at the heart of agricultural systems. The spread of large scale commercial agriculture has put family unit agriculture under stress. Efforts are needed to ensure that technology innovations and other changes support sustainable family agriculture (Francis, 1994).

In many countries there is increasing utilization of marginal, often harsh, environments as population increases and capitalism spreads. Marginal land agriculture by both poorer farmers and commercial agriculture is often based on rain-fed cultivation, but this is likely to lead to unsustainable yields, failed livelihoods and environmental degradation. Runoff agriculture should offer more security, better yields and more sustainability than rain-fed. However, knowledge of a region's ecology and a repertoire of potentially successful techniques does not guarantee adequate, sustained agricultural production. Should any livelihood problems arise smallholders tend to maximize short-term gains – if need be at the expense of sustainable production – so efforts at innovation must quickly 'get it right' (Collins, 1986).

Sometimes the breakdown of indigenous practices may not matter if there are better alternatives; Pereira (1989, p168) suggested that it might not be too bad if traditional *quanat* systems declined because cheap plastic pipes and motor pumps now mean that the labour input is uneconomic. Dwindling *quanat* systems allow a dangerous and degrading system of labour (for maintaining and building *quanats*) to disappear. However, there are often no better alternatives available, and the shift is to something environmentally damaging, or to rural–urban migration. Worldwide mainstream irrigation has attracted investment, while SWC and runoff agriculture has not. Commerce can clearly see less long-term profit to be made in approaches that involve few outside inputs!

Traditions which ensured or encouraged community cooperation, group labour and satisfactory management of common resources are increasingly

BOX 5.1 RUNOFF AGRICULTURE

In a given situation runoff agriculture may be affected by one or more of the following:

- population changes;
- social change (collapse of traditional authority, rules or cooperation);
- subdivision of landholdings;
- practitioners not adequately benefitting;
- resentment of top-down extension;
- difficulty getting materials for maintenance;
- returns for labour change (if returns for labour invested in runoff agriculture are not attractive there will be neglect or damaging overexploitation);
- market changes affecting sale of produce;
- globalization, trade agreements, structural adjustment programmes;
- agricultural policies (eg reduction of or fewer subsidized inputs);
- decline of community self-management, often due to bureaucratic intervention by government, or because labourers are attracted away by employment opportunities in urban areas, tourism or overseas;
- technological innovation, such as the spread of ploughs, motor pumps, tube wells;
- insecurity of land tenure;
- labour migration (spontaneous or encouraged by state; migration may lead to breakdown, but in some cases may provide investment for improvement or expansion;
- risk perception;
- appearance of pest organisms, crop or human diseases;
- insecurity and unrest (runoff agriculture may fail if farmers are disturbed by civil unrest, but there are also cases where people have abandoned runoff agriculture and moved out of highlands once danger passes);
- changes in communications (this may open up opportunities or let in cheaper competition and damage market prices);
- pollution (such as acid deposition);
- global climate change;
- import of land degradation higher in the catchment (eg overgrazing or logging removes tree or brush cover and leads to soil erosion, less capture of rainfall or snow, which in turn reduces streamflow and causes springs to dry up.
- Introduction of new crops, tchniques or controls which offer opportunities to improve runoff cultivation and/or halt degradation of higher areas of catchment.

under stress. Finding alternatives to traditional respect for authority and community cooperation is a widespread problem that must be overcome before runoff agriculture is likely to be successful.

If there is to be no breakdown, SWC and runoff agriculture must be carefully planned and structures must be well constructed and well managed. It is important to understand why people seeking to sustain an adequate livelihood may degrade the land and other resources, and why they find it difficult to shift to less damaging practices. Various researchers have addressed the problem of land degradation – notably Blaikie, who explored the political economy of soil erosion (Blaikie, 1985), and a growing number are studying the process and control of land degradation (Blaikie and Brookfield, 1987; Little and Horrowitz, 1987; Johnson and Lewis, 1995). Much of this research is relevant to the breakdown of traditional livelihood strategies, including the abandonment or degeneration of runoff agriculture.

Sometimes the cause of breakdown of livelihood strategy is easily identified – for example, loss of rural labour through migration or excessive subdivision of landholdings. However, generalization is difficult and unwise, runoff agriculture strategies are frequently location-specific and the challenges leading to difficulties and breakdown are diverse, often insidious, and vary from region to region, although there may be some that are commonly shared. Furthermore, strategies may fail because of a complex of reasons, rather than one simple cause. In some situations the cause and effect process is so complex it is difficult to separate them from each other – for example, did drought drive farmers to migrate, or, having lost labour through migration, did farmers neglect terracing (or other labour-demanding strategies) and so suffer more when drought struck? Some problems are clearly apparent, for instance, when SWC has been carried out using mechanization and local people find it difficult to maintain structures with hand tools and cannot afford or get access to appropriate equipment (Benasalem, 1981). Successful transfer of those runoff agriculture methods which are flourishing in one region to another can be problematic.

Planning runoff agriculture requires data; its gathering should ideally involve mutual trust and collaborative contact between local people and planners for an adequate period of time. In practice, this may be difficult, especially where people are suspicious of outsiders and when there is little time available for development to take place. Such a collaborative approach to data-gathering was adopted by Garcia-Perez et al (1995) when researching indigenous terrace breakdown in Spain. Others have adopted a collaborative approach; for example, Reeve and Black (1994) tried to classify farmers' attitudes to land degradation as a preparation for designing SWC strategies.

The returns from planned SWC and runoff agriculture development must be carefully assessed before proceeding, and it is vital to establish what will motivate farmers. These may be difficult to determine when SWC and runoff agriculture are part of a complex livelihood strategy. Diversification of crops and risk reduction (improving the chances of a successful harvest) may be more important to farmers than immediate and obvious yield improvements or long-term sustainability (the latter may be especially difficult to sell to smallholders). Given adequate security of livelihood, smallfarmers will experiment and adopt new innovations. However, there is a need for more research on these issues. Van Dijk (1997) examined the reasons for the periodic expansion and contraction of indigenous SWC in eastern Sudan since the late 1970s. He found that there was village to village and year to year variation and concluded that historical research was important for unravelling the often complex causes of decline or expansion of SWC or runoff agriculture. In eastern Sudan SWC (in the form of *teras* – see Chapter 3) appear to diversify incomes in normal years and in difficult years offer real advantages as a means of reducing risk of livelihood failure; however, this was not immediately obvious to observers.

A challenge to those developing SWC and runoff agriculture is to develop approaches which help counter *dry* soil erosion as well as runoff erosion. Activities such as hand hoeing can lead to serious soil erosion when it is dry. SWC may therefore fail to halt land degradation or to sustain production because dry period activities are overlooked. There may even be situations where a change to SWC encourages tillage practices which lead to dry soil damage. Studies in Rwanda by Lewis and Nyamulinda (1996) indicate that soil movements due to hoeing in dry periods were a considerably greater cause of land damage than runoff. Countermeasures may be needed in addition to adequate SWC to cope with tillage – for example, vegetative barriers between terraces.

WHERE HAS BREAKDOWN OCCURRED?

An introduction to the history of SWC and runoff agriculture can be found in the journal *Agricultural History*, vol 59(2) pp103–372, published in 1985.

Latin America

The greatest loss of runoff agriculture has probably taken place in South and Central America since the Spanish Conquest (Donkin, 1979; Matheny, 1982; Park, 1983). In Peru alone, a large part of approximately one million hectares of terracing has been lost (Guillet, 1987). The conquest led to depopulation caused by introduced diseases and dislocation of populations, together with the collapse of effective social institutions and morale, and the introduction of debt-peonage (a form of near slavery). Terracing and other forms of SWC flourished from about 500 BC until the arrival of the Spanish; however, there were phases of expansion and abandonment during that time, possibly reflecting periods of better precipitation or family growth and ageing (Devevan, 1986; Guillet, 1987).

More recently, surviving terrace agriculture has often suffered as a consequence of peasant farmers seeking non-farm employment. The result of this shift is either abandonment or part-time farming, both of which lead to neglect of structures and land degradation (Zimmerer, 1994). Modern terracing in Latin America is often poorly constructed and managed, compared with past efforts. Archaeology reveals techniques that might work effectively today to give better crops and more sustained production than current practices; the problem is to improve motivation, management and investment (Farrington, 1980; Beach and Dunning, 1995).

In many parts of the world problems are caused for runoff agriculture by geomorphological or tectonic factors. Landslides, avalanches and earthquakes can often be avoided if adequate hazard mapping is done and the results heeded. Tectonic uplift has been a problem in coastal Peru and Chile and in other parts of the Andes where it may damage channels, alter the gradient of terraces and cause gullying and slumping (Ortloff, 1988).

North America

There is a tradition of runoff agriculture in south-western US, and some native peoples still practice these methods (see Chapter 4). However, much has been lost over the last few centuries (Sandar et al, 1990); diseases caught in the early days of European settlement are probably one of the main causes of decline, with whole tribes being decimated.

Israel (Negev Desert)

The Nabateans practised runoff agriculture in the Negev between about 300 BC and AD 650; control of important caravan trade routes and livestock herding also supported their settlements. Socio-political changes from the middle of the 7th century AD led to a gradual decline of sedentary agriculture. People became more nomadic and only the most easily managed terraces and cisterns remained in use (Johnson and Lewis, 1995, pp28–41)

Lebanon

There are widespread examples of ancient, abandoned runoff agriculture systems in the Lebanon; some may well have potential for renovation (Zurayk, 1994).

Yemen

Highland areas of Yemen have a tradition of irrigated terrace agriculture and runoff harvesting dating back many centuries, some of it still functioning (Varisco, 1983). In the Yemen the Sabean people built large earth bunds across *wadis* to divert runoff to croplands. During the 19th century AD, after around 2000 years of operation, the bunds became flood damaged and were not maintained, probably as a consequence of losing powerful political leaders and the ability to organize society (Bunner, 1986). From about 1971 there has been considerable neglect and abandonment as large numbers of adult males migrated to the Gulf states to seek employment (Vogel, 1985; 1993).

Spain

There have been a number of recent studies of terrace neglect and failure in Spain. The Moors established terraced agriculture in Spain in the 15th century and many of their systems remained in use until the 19th century with little breakdown. Rural depopulation and market price fluctuations for crops, notably almonds, led to terrace abandonment in some areas. In parts of southern Spain the *seccano* system of runoff harvesting terraces (non-irrigated terraces) has suffered for the last 40 years mainly, it seems, as a result of outmigration to urban or coastal tourism-related employment. The decline has been accelerated by European Union (EU) agricultural policy and EU social policy, together with a shift to more ownership by absentees and less pride in farming as opposed to profit motives (Douglas et al, 1994; Thompson and Scoging, 1995; Rodriguez-Aizpeoplea and Lasanta Martinez, 1992).

In recent years EU agricultural policy has encouraged expanded cultivation of certain crops (for instance, cereals and almonds). Subsidies have been on a per hectare basis and have apparently undermined traditional good management of smallholdings. Hasty clearance of scrub for cereal or treecrop cultivation to take advantage of subsidies, often on steep slopes with unsuitable soils, and no SWC practices or poorly constructed, mechanically dug terracing, has led to gully erosion and other forms of land degradation (Faulkner, 1995).

Madeira

Madeira is self-supporting in fruit and vegetables and produces considerable quantities of wine and bananas for export; most of this produce comes from small, irrigated terraces farmed mainly by hand on very steep slopes. The intensive, largely sustainable runoff agriculture sector has come under threat from restrictions on banana exports, which means that alternative export crops may need to be found. Outmigration of younger people has not helped the maintenance of terraces. Furthermore, the expansion of tourism and road construction has led to escalating land values and easier access to tourism employment; consequently, land is being abandoned and sold for house building. Somehow non-viable smallholdings must be amalgamated and returns from terrace farming improved to raise rural incomes.

Canary Islands

The islands have considerable areas of terrace agriculture, irrigated by diversion of springs, seasonal streams and runoff water stored in cisterns. Problems have been caused by deforestation of upper slopes in Tenerife and there is growing competition for water supplies from tourism. In some parts of the islands groundwater has become contaminated with salts and agrochemicals as a consequence of intensive irrigated agriculture, which may necessitate more use of runoff water to avoid soil salinization. Reafforestation should help to trap mist and recharge groundwater and improve the flow of streams and springs.

Cape Verde Islands

A long history of exploitative farming to produce export crops of sugar and other products, as well as overgrazing and fuelwood collection, has led to widespread soil degradation and reduced groundwater recharge. The situation was recently reviewed by Langworthy and Finan (1997), who made an ethnographic study of Cape Verde's struggle for self-sufficient food production, which has to be balanced against a more sustainable use of the environment. These studies include an examination of indigenous approaches to landuse. Some of the strategies which have functioned for long periods, in some cases centuries, have failed or are starting to break down because males of working age have left to seek employment or wives in greater numbers since the 1960s (Hallsworth, 1997).

Agriculture in the Cape Verde islands is difficult without terraces for soil and water conservation and to provide level plots on the steep slopes. The islands have suffered from droughts nearly as badly as the Sahel since the 1960s. Outmigration has seriously reduced the availability of labour to maintain terraces (IFAD, 1992, p22). The authorities have adopted a work-for-relief programme (*frentes do trabalho*) to extend SWC and runoff agriculture measures, with – at best – mixed results. Those interested in government support for runoff agriculture or SWC extension should examine the Cape Verde experience with environmental public works. Often landowners who have received *frentes do trabalho* support fail to effectively use or maintain structures. The reasons are complex, but in some part this results from a top-down approach by the authorities, which means *frentes do trabalho* works are often mis-sited or provided for those who do not really want them.

Aegean Islands

The Aegean Islands have a tradition of terraced agriculture, mainly depending on runoff collection for moisture and producing olives, figs, almonds, vines and annual crops. Since the 1950s tractors have been increasingly adopted, and farmers have found it more profitable to grow annual crops, such as sunflowers, in valley bottoms and lowlands. There has been neglect of terracing in many areas as a consequence, but the development of intensive agriculture on the plains of mainland Greece, where economies of scale and mechanization are possible has also been a contributing factor; this has helped lower the prices for agricultural products traditionally grown by terraced cultivation. Terrace-grown olives and dried fruit now offer little profit. The result has been the abandonment of terraces and the collapse of traditional agropastoral systems. Deprived of terraced land income, families often increase livestock grazing of higher altitude scrubland and forests; as a result, land degradation follows and groundwater recharge may suffer (Lehmann, 1993; Margaris, 1992).

Another factor that has driven down market prices of terraced agriculture produce has been the change in consumption and cooking habits of European households, notably the shift to use fresh fruit shipped from New Zealand, Chile and South Africa (Margaris, 1987). Working women have less time to prepare meals, tastes have changed, and consumption of dried fruit and other traditional terrace agriculture crops has declined. Current fashion leading to consumption of more natural foods may help to check this decline, but the availability of dried fruit from the US, South America, South Africa, where mechanization, cheap labour and other advantages lower costs below those of the Aegean Islands, does not help. Also, many consumers have moved to frozen foods rather than dried fruit.

Other Mediterranean Islands

Malta and the nearby island of Gozo still have extensive systems of (non-irrigated, runoff-collecting) terracing which produce wine and agricultural produce even though annual rainfall seldom exceeds 600 millimetres per year. Abandonment and neglect is increasing since working age people tend to seek employment in urban areas, the tourism sector or off the island.

India and Sri Lanka

Some of the problems of tank irrigation since colonial times have already been outlined in Chapter 4. The degeneration and breakdown of runoff storage using tanks has attracted considerable interest, some seeking to ascertain why, others trying to identify cures and improvements (Govindasamy and Balasubramaniam, 1990; Shankari, 1991; Reddy et al, 1993). In general, problems have been associated with a relaxation of feudal landlord–peasant farmer relationships. This breakdown has led to reduced maintenance since farmers no longer feel obligated to clear tanks and use the silt on their land. The costs of clearing and spreading silt are also unattractive now that chemical fertilizers are available. There are yet more problems; rising population often means more agriculture in upper parts of catchments and this leads to greater siltation of tanks. There is also a growing practice of illegal cultivation of partly silted-up tanks, so silt removal is neglected, tanks are damaged by wet season flows and land degrades. The solution might be more widespread use of partic
ipatory management approaches (Ambler, 1994).

The Philippines

Spectacular and ancient irrigated rice terraces have been cultivated in the Banaue district for over 2000 years. Recently these have become damaged by the accidental introduction of 'giant' earthworms. Up to 36 centimetres in length, these worms probably arrived with new rice plants or agricultural equipment brought in to improve yields. Disastrously, they bore into terrace retaining walls and cause progressive collapse (*The Times*, 29 November 1997, p30). In addition to these difficulties, the maintenance of these terraces has also been hit by agricultural stagnation and social change (Eder, 1982).

Pakistan

Since the 1960s tribal organizations and individuals have apparently been encouraged by government grants and loans to install tube wells. In many areas tube wells have lowered the watertable causing *karez* (*quanat*) systems to fail. The tube wells which replace the *karez* systems are probably not as sustainable or as equitable a strategy. The tendency is for groundwater to be overexploited by tube wells so that it fails, whereas *karez* systems put groundwater under less strain. There have also been problems maintaining *karez* systems: the traditional employment of digging them has suffered because there are opportunities to earn better money elsewhere. In the past farmers or villagers invested in *karez* systems and enjoyed water rights related to their investment; recently investment has shifted to other things, especially urban business or transport (Khan, 1995).

Iran

There is a very long history of agriculture in the Fertile Crescent, which includes Iran, as well as runoff exploitation, especially through the *quanat*. A number of researchers have examined the breakdown of Iran's very numerous *quanat* systems in the second half of the 20th century; one is Bonine (1996), who provides information on importance, construction costs and trends. The major causes of breakdown seem to be social change, the penetration of capitalism (see Chapter 3) and the spread of tube well and motor pump technology.

Eastern Sudan

Teras are a traditional SWC strategy of the eastern Sudan consisting of earthen bunds which trap overland flow (see Chapter 3). Studies have indicated that there has been abandonment in some regions and expansion in others. These trends do not seem to be responses to drought. Van Dijk (1997) concluded that they reflect varying challenges to overall livelihood strategies. It seems that local people weigh labour on *teras* against benefits from labour migration to off-farm employment or other activities making up household income. The *teras* offered as much as 31 per cent of total household livelihood in 1997, and in drought years help reduce risks of livelihood failure.

Tanzania

The *wa matengo* pit system of SWC (see Chapter 2) has suffered, possibly because of labour losses through male migration to urban areas (Hallsworth, 1987, p22).

Niger

SWC and runoff agriculture appear to have undergone phases of expansion and contraction as markets change, employment opportunities appear and

(C J Barrow, 1998)

Photograph 5.1 Forest degradation in the High Atlas Mountains, Morocco.
Eroding soil chokes channels and damages terraces at lower altitude.
Loss of vegetation cover may be to blame for recent poor retention of
winter snow which has reduced summer streamflow, endangering
irrigated terrace cultivation in some areas

decline, and as family attitudes and social institutions alter. Here, and in some
other parts of sub-Saharan Africa, there have been recent signs of increased
use of stone lines and pit systems after a period of neglect (Reij et al, 1988,
pp20–22). Similar shifts from decline to expansion and vice versa have been
noted elsewhere, for example in South Yemen. It is therefore important for
extension services to assess trends before trying to promote SWC.

Morocco

Morocco has a rich tradition of runoff agriculture which has suffered in recent
decades as a consequence of national agricultural and economic policies,
penetration of capital into rural areas, social change, and population growth
(Kutsch, 1982). Irrigated terrace agriculture in the Rif, Middle and High Atlas
mountains has suffered in some areas because of population increase and
subdivision of holdings, low market prices for produce, poor communications,
and changing attitudes toward village management and social obligations.
Profits to be made from runoff agriculture may fail to keep farmers on the
land. Furthermore, some leave to find off-farm employment to save for

(C J Barrow, 1998)

Photograph 5.2 Forested a few years ago, this area of the High Atlas Mountains in Morocco is now of little use other than for rough grazing. Abandoned stone-lines and terraces and worsening gullying show that there has been some abandonment of runoff cultivation

marriage and to improve their agriculture in the future. Labour shortage sooner or later leads to neglect of SWC and land degradation. Terrace structures fail through poor maintenance or because those trying to maintain an adequate livelihood from agriculture expand livestock herding when runoff agriculture yields little income. Overgrazing, shifting cultivation and excessive fuelwood collection in uplands has led to land degradation, which can damage runoff agriculture and have impacts further afield in the lowlands (Ait Hamza, 1996; Chaker et al, 1996).

The traditional Moroccan *khettara* (*quanat*) system has also suffered from social change, off-farm employment and the acquisition of pump sets by richer landowners (Lightfoot, 1996). State support has often been directed at large-scale commercial agriculture and has done little to aid traditional cultivation. The result tends to be highland degradation, which damages small- and large scale, lowland irrigation schemes and urban water supplies, and can generate a flow of farmers to the cities seeking employment, of which there is little to be had (Photographs 5.1 and 5.2). The neglect of SWC and small scale runoff agriculture can thus have far-reaching socio-economic impacts which range beyond farmlands into the cities of Morocco and might even (as a consequence of migration) affect other Maghreb nations and Europe. It will be interesting to see whether the effects of structural adjustment, especially in reducing chances of employment for urban migrants, may finally stimulate a renewed interest in runoff agriculture.

Tunisia

The *meskat* runoff harvesting systems of Tunisia, having declined from considerable importance in Roman times, have suffered further breakdown since the 1970s, largely because of increasing demand for building land which reduces the size of catchment areas and makes the strategy non viable (Reij et al, 1988, p20). Problems also seem to have been caused by land reform, which has led to young and dynamic people migrating off the land. Excessive olive planting has been taking place in some areas (whether this is a consequence of labour shortages caused by rural-urban migration is difficult to establish). This leads to greater moisture use which, combined with lack of manpower for maintenance of runoff agriculture, leads to breakdowns. In Tunisia, and probably other Maghreb countries, indigenous methods remind people of feudalism and past colonial conditions and this probably helps encourage abandonment and resistance to reviving established methods. The mid to late Roman decline of Mediterranean intensive agriculture has been blamed on rising labour costs, partly caused by reduced use of slave labour.

6 EXPANDING, UPGRADING AND REHABILITATING RUNOFF AGRICULTURE

The changes in English agriculture grouped by historians under the heading 'the Agricultural Revolution' were brought about by farmers not scientists (Richards, 1985, p117).

One of the most valuable aspects of runoff agriculture and SWC is that both offer possibilities for sustainable rural livelihoods that can be grasped by small-holders without the necessity for large amounts of credit, expensive inputs or infrastructure. Often the poor do not gain from innovation, while the rich do – with runoff agriculture this might be avoided (Critchley et al 1992). Efforts to upgrade agriculture can lead to significant inequalities between those in favoured localities, and in receipt of support, and others in less favourable areas who have no aid. Runoff agriculture and SWC may be a way of reducing this sort of difficulty: both should be accessible to the poorest of agricultural-ists, even in quite remote and harsh environments.

Agricultural development has tended to concentrate on yield improve-ment (especially working with the large-scale, commercial sector) and not on sustaining soil and water inputs and ensuring equitable, secure socio-economic development. Runoff agriculture and SWC have the potential to improve security of harvests, should improve yields, and might well be more accessible and equitable. It should also be more sustainable, do less off-site environmental damage and lead to less dependency. Therefore, runoff agricul-ture and SWC have a number of advantages; furthermore, in many situations (especially marginal environments) there are unlikely to be practical, accessi-ble alternatives in the foreseeable future.

Runoff agriculture development can be focused on low-input approaches suitable for small scale agriculture and can promote good land husbandry, especially the adoption of effective SWC (Gischler and Fernandes Jauregui, 1984; Lal, 1990); it could also be more widely adopted by large scale agricul-ture. Agroindustrial investors have not shown much interest, probably because there are limited opportunities for sales of, for example, agrochemicals and pumps. Large growers may also be reluctant to adopt runoff agriculture because the yield increases are limited and sustainability is undervalued. There may be possibilities for small farmers to adopt runoff agriculture and to become involved in *contract growing*. This involves a large company or cooperative coordinating transport and marketing of produce that is supplied by contracted growers – sometimes smallholders – to strict schedules and standards and preagreed prices. The growers know what price they will get well in advance, even if crops or the market are poor, but cannot expect peak market price if the commodity attracts it. They are also assisted in breaking into difficult markets. Arrangements already exist where small farmers in Africa and South East Asia produce luxury vegetables for developed country supermarket chains.

The goal of those adopting runoff agriculture and SWC is often to survive and to counter land degradation, drought and famine or the breakdown of established landuse strategies. Before embarking on any promotion of runoff agriculture, efforts should be made to assess whether it might be better to attract agriculturalists to less fragile environments or towards non-agricultural employment opportunities; although ruoff agriculture has great potential, relocation or re-employment outside of agriculture may sometimes offer less of a challenge and more chance of sustained livelihood and reduced environmental damage (Forster, 1992).

Efforts to upgrade agriculture and conserve soil have until recently tended to ignore indigenous knowledge and techniques (Haagsma, 1995); however, these have great potential for improving agriculture and natural resources management (De Walt, 1994). Agricultural extension and SWC has often been top down, relying on sanctions or penalties, and sometimes even based on forced labour. The result is that, even where pre-independence SWC and runoff agriculture were conducted in a reasonably sensitive and effective way, they became a reminder of colonial times and so were often rejected. Decades after independence, people ignored or resisted SWC and runoff agriculture. For example, in Rwanda and Burundi, Belgian use of forced labour made people hostile to later attempts at much needed SWC (Anderson, 1984; Critchley et al, 1992, p27).

Runoff agriculture and SWC efforts have sometimes been unsustained and, in some cases, have failed, causing serious environmental degradation and human hardship. There has been a tendency for authorities, and even some NGOs, to adopt insensitive approaches to upgrading or expanding runoff agriculture and SWC. This has taken the form of promoting mechanized construction of SWC or runoff agriculture structures when users may not have good enough access to machinery to repair and maintain the structures. Structures have often been poorly designed, installed in the wrong type of site, or are of poor workmanship. Belsky (1994) examined SWC efforts in upland Sumatra since the colonial era, noting the emphasis on contour bench terraces, which might be suitable for commercially orientated agriculture but not for smallfarmers; this imposition of inappropriate mechanical terracing led to a decline in diversity of the upland farming systems. Attempts to promote SWC and runoff agriculture have occasionally resulted in crop yields little better, and sometimes much worse, than previous practices (Odemerho and

Avwunudiogba, 1993). This can have far-reaching effects; when peasant farmers see innovation fail, they are likely to treat future efforts with suspicion and the lesson spreads much wider, even to other regions.

SWC and runoff agriculture may be technically feasible and appear simple, only demanding locally available materials – but that may not be enough; innovation must 'fit' environmental and socio-economic conditions and appeal to farmers (Jurion and Henry, 1969; Renner and Frasier, 1995). Indigenous tenurial rights and farmers' obligations can be complex and difficult for outsiders to unravel. Often such rights and obligations and access to common resources are being lost as modernization and capital penetration occurs. There must also be resources to back innovation, adequate assessment and monitoring to 'steer' things, and clearly prioritized goals (Taabni and Kouti, 1993). To make runoff agriculture and SWC sustainable may not be easy but it must be a goal (Pretty and Shah, 1997). Approaches such as *farming systems research* (FSR) (see Glossary) offer promising ways of ensuring innovation is appropriate and will be supported by farmers.

Caution is required when exploring the economics of SWC and runoff agriculture; there may be off-farm benefits that are difficult to assess, and in challenging environments, where landusers exist at or near the margin, only a very little extra soil moisture or water stored in a cistern may make the difference between survival and failure – conventional economics do not apply. Often improved security of harvest outweighs any increase in yields as far as small scale landusers are concerned, and planners may not adequately appreciate this.

There has been increased interest in SWC and runoff agriculture, especially in sub-Saharan Africa, probably as a consequence of 1970s and 1980s droughts. However, efforts to promote indigenous SWC and runoff agriculture have had patchy success (Reij et al, 1986; Critchley et al, 1994). For example, *fanya juu* terraces have been successfully spread in some parts of the Kenyan highlands, but failed to take off in some of Kenya's semiarid areas (Reij et al, 1988, p20). Pretty and Critchley (1997) noted that the history of SWC and runoff agriculture extension has been one of landusers being advised, paid or forced to adopt new measures and practices. They call for a new era where sustainability is valued and the benefits of local knowledge and skills are put at the heart of appropriate developments, which are supported by the people involved. The challenge is to spread traditional methods which work, to help establish them, and to build upon them without losing their strengths – the emphasis is on local adaption and participation (Critchley, 1989; Adams, 1990, p1320), what Cullis and Pacey (1992) termed 'interactive technology development'. Other development experts have stressed the need for more aid to be focused on helping resourceful and innovative cultivators, rather than supporting those who are failing. Broadly, one may say that farmers will adopt innovations that they find attractive: which they feel work. They may do so even with little or no help; unattractive innovations will be ignored no matter how much and how good the extension aid given. It therefore makes sense to check early on whether innovations are attractive to farmers.

Being an attractive innovation is not enough; the reasons why efforts to extend or upgrade SWC and runoff agriculture fail are numerous (see Box 6.1).

Box 6.1 Failures in Extending or Upgrading Runoff Agriculture

Reasons why efforts to extend or upgrade runoff agriculture and SWC fail may include the following (more than one of these may operate in a given situation):

- locals do not really profit (from SWC) and benefits may accrue to people off-farm – for example, reduced erosion (achieved by farmers' efforts) benefits water users elsewhere by reducing siltation or recharging groundwater, but does little for the farmer who expended effort;
- farmers are not adequately consulted or involved and become resentful;
- there is an emphasis on engineering and/or new crops, leading to neglect of crucial factors, such as transport, marketing, inputs and land tenure;
- labour availability is overestimated and innovations cannot be supported;
- the role of women missed – for example, women are unable to retain benefits once they adopt and manage SWC or runoff cultivation, or SWC involves more labour than they can manage, or menfolk dominate supply of inputs or marketing;
- poor design;
- poor siting;
- poor construction;
- inappropriate techniques which locals cannot maintain;
- some stress the need for local people to be empowered, to ensure success; this may imply some degree of decentralization of decision-making and control (which can make response to problems of day-to-day management quicker);
- the approach is unsuited to the local environment;
- failure to address problem of overpopulation: 'no amount of terracing will solve the problem of overpopulation' (Hilsum, 1992, p5);
- incorrect estimates of likely annual rainfall; reliance on mean estimates can be dangerous when precipitation varies from year to year;
- aid or grants act as a disincentive to self-reliance (Grepperud, 1995);
- poor extension and training of users;
- benefits are long-term but users seek and need short-term gains (financial or labour saving or improved security of livelihood) (Jolly et al, 1985; Swanson, et al, 1986);
- efforts are negated by unforeseen forces (such as social unrest, diseases and new fashions);
- efforts lead to a regional market glut for a given crop; if there is no good storage, transport or way of getting a satisfactory return there is little point in improvements.

There is a huge diversity of indigenous SWC and runoff agriculture strategies. These have enormous potential to extend to new areas and to form the basis for new approaches. There are parallels in extending or modernizing traditional irrigation methods, small scale agriculture, SWC and runoff agriculture (Mock, 1985; Adams, 1989; 1990; Kerr and Sanghi, 1992; Van der Waal and Zaal, 1990; Van der Ploeg and Van Dijk, 1995). However, although there are lessons that can be shared, experience shows that approaches such as SWC and runoff agriculture do not always transfer well from one situation to another (Warren, 1991; Warren et al, 1998); therefore, traditional approaches need careful evaluation (Moussa, 1997).

It is important that those seeking development do not impose 'contemporary thinking about development' on the landscape and local people (Adams, 1996, p156). For example, it has become fashionable in some quarters to reject large scale developments in favour of decentralized, smaller-scale approaches; indigenous approaches have also become attractive. There may, however, be situations where large scale, centralized development is appropriate – developers must ignore fashion and promote what works, is sustainable and is appropriate (Groenfeldt, 1991). It might also be wise to seek a mix of strategies so that if one approach fails in future, others will still function.

Some indigenous strategies have attracted the attention of modern agricultural improvement researchers – for instance, the run-on farming of the Negev, terracing in Morocco, Libya and the Yemen, and the raised fields and swamp cultivation of Latin America. Some of these may no longer be viable because of environmental and socio-economic change (such as salinization, climatic change and soil structural changes), rural depopulation, or the loss of very cheap labourers or even the existence of slave labour. Modern techniques may overcome some of those problems and resurrect them (Zurayak, 1994). Rajaram et al (1991) compared indigenous and established modern tillage systems for their effectiveness in sustainable food production – modern, mechanized tillage, they concluded, had much to learn from traditional methods.

RUNOFF HARVESTING

Since the early 1960s there has been a good deal of interest in spreading runoff harvesting (see Chapter 3 for further details of strategies and techniques), and there has been some success, especially in sub-Saharan Africa (van Dijk and Ahmed, 1993). Rapp and Hasteen-Dahlin (1990) review the promotion of water harvesting for the drylands of developing countries, while Rapp and Frasier (1995) examine the socio-economic design elements.

FLOOD AND WETLAND AGRICULTURE

Flood and wetland agriculture provide a livelihood for a great many small-farmers and in some countries yield the bulk of food supplies (flood and wetland agriculture is discussed in Chapter 4). Development efforts have done little to help this sector but have continued to support large dams and barrages, mainstream irrigation schemes and industry, all of which frequently disrupt wetland and floodland environments, wreaking havock on wildlife and on indigenous landusers. Large-scale developments have become relatively more costly, and because they often generate so many impacts, it is high time to study and invest in cheaper, more appropriate and sustainable flood and wetland agriculture.

Tank Storage

In South Asia, especially the regions subject to seasonal monsoon rainfall, tanks have long been used as a way for communities to store surplus runoff for agricultural use during dry seasons, and for domestic supply and fish production. For a range of socio-economic reasons (see Chapter 5), tank construction and maintenance have in many areas fallen into neglect, prompting efforts to rehabilitate and stimulate construction of new systems (Murray-Rust and Rao, 1987).

In the dry zone of Sri Lanka, tank storage expanded between 1950 and 1970 and was mainly achieved by improving organizational support through the creation of tank committees, designed to meet regularly and bring together government officers and farmers' leaders to discuss needs, plans and progress. Much was achieved through the Tank Irrigation Modernization Project, Sri Lanka's first major effort at rehabilitation and upgrading, during which the irrigation department carried out construction work funded by the World Bank and the Overseas Development Institute (ODI) of the UK. This project demonstrated the need for operational and institutional efforts to be well managed; it also made it clear that developers must work with local people (in southern India there is often close association between village temple and tanks) (Kasivelu et al, 1995, p49) and that maintenance funding must be satisfactory, which in the long run probably requires improved collection of service fees from water users (Murray-Rust, 1987).

Non-Sedentary Agriculture

Non-sedentary agriculture (shifting cultivation or bush fallow, and nomadic pastoralism) is often in a state of decline, causing poverty and environmental degradation. Where human population is increasing, a point is reached when non-sedentary agriculture is unable to find enough land to ensure that a plot recovers before subsequent reuse. For some environments there must be enough land to allow 20 or more years fallow (so that for every hectare cropped, 20 or more hectares must remain as set-aside). In addition to population increase, shifting cultivation may degrade because people have been forced to settle land of poor quality, or because market opportunities encourage the growing of speculative crops, often at remote sites (such as narcotics).

A large literature has been generated on the problems of shifting agriculture and how it might be replaced by productive and sustainable systems that are able to support larger populations. The problem is that shifting cultivation is often practised in difficult conditions and modern agriculture may not be able to offer anything better, or even as good. Furthermore, the shifting cultivator is likely to be poor and cut off from outside communication, so any assistance may give little cash return. Many who practise shifting agriculture are 'shifted agriculturalists' – people forced to move by war, persecution or resettlement. They may have few tools or resources, or perhaps little hope of returning to their tradition of sedentary agriculture, or are disorientated by failed land settlement efforts and therefore struggle to survive with poorly managed shifting cultivation. There may be problems converting non-sedentary agriculturalists to sedentary agriculture if there is no tradition of an annual farming calendar or forward planning. Developing country administrators often hold non-sedentary agriculture in contempt, attributing to it the stigma of backwardness; those involved are often weak and poor, and often bureaucrats wish to see those practising it 'sedentarized' so that they can be policed and controlled. In the past, administrators often saw the diversified cropping strategies of shifting agriculturalists as ill organized and inferior. For these reasons, the strengths of non-sedentary strategies were long overlooked, though they have much to offer to modern agriculture.

The problem is how can shifting agriculture give better yields when there is insufficient land to allow adequate land rotation? Yields might be improved, and soil degradation reduced after a plot is cleared, if runoff agriculture and SWC can be tailored to fit shifting cultivators' needs.

PROMOTING RUNOFF AGRICULTURE AND SOIL AND WATER CONSERVATION

It is important to assess what land would benefit from SWC or is suitable for runoff agriculture. GIS and land capability survey can be useful, although socio-economic factors need to be considered alongside the biogeophysical (Taller et al, 1991; Tauer and Humborg, 1992). Making resource surveys and land capability maps is not enough; ministries may fail to make them available and without effective policies little will be achieved.

Small-scale farmers are experts on local conditions and what they do usually works, not least because there is unlikely to be any welfare system available to them if they fail. Given these realities, they are cautious and are also constrained by many other difficulties. Nevertheless, they can be 'dynamic and innovative' (Richards, 1985, p14), and the promotion of SWC and runoff agriculture can be successful. Chances of success are improved if extension services understand the livelihood strategies that are in place and those which they hope to promote.

Livelihood strategies are seldom simple and might be better understood and managed if their component parts are examined. In the case of SWC and runoff agriculture, they can be split into:

- environmental and technical aspects;
- social and economic aspects;
- institutional and policy aspects (Critchley et al, 1992, p39).

Box 6.2 lists points which need to be considered before promoting SWC and runoff agriculture. Ideally, any such efforts should proceed after a properly studied pilot scheme. It is also useful to explore how innovations will interact with existing and planned programmes and policies (techniques of strategic impact assessment could be used). In practice, pilot studies are too seldom used and the strategic assessment of projects, programmes and policies is still new.

BOX 6.2 PROMOTING RUNOFF AGRICULTURE OR SWC

Points which should be considered before attempting to promote runoff agriculture or SWC include the following.

- At the outset, anyone considering the promotion of runoff agriculture or SWC must ask: do the likely results justify the costs?
- Can agriculture be improved without runoff agriculture or SWC (McCown et al, 1992) – for instance, by making modest amounts of chemical fertilizer available or introducing green manuring?
- Technology may work on an experimental farm or research station, but it may not function effectively in the real world; the developer must ask whether it suits the needs and capabilities of local people and fits the environment.
- Ideally, runoff agriculture or SWC will improve sustainability of landuse, but that alone is unlikely to prompt adoption (see the next point).
- Runoff agriculture or SWC alone will not motivate most landusers (whether developed country farmers or developing country smallholders); it must be integrated with agricultural development. If runoff agriculture or SWC conserve soil *and* improve crops or security of harvest, or provide fuelwood, they are more likely to get support.
- Is there the political will (Hudson, 1987, p5) to make runoff agriculture or SWC work?
- It is important to plan for the unexpected and to have good safety margins if strategies are to be sustainable.
- Soil erosion may occur during a very limited period – for example, just after ploughing (Hilsum, 1992, p3), or during occasional storms; what is the best runoff agriculture to address this?
- There may not be much sense in terracing if livestock is uncontrolled and causes damage.
- If the state levies too much tax on harvests, farmers will not be able to reinvest some of their profits in land husbandry; however, so innovation is probably doomed.
- Failure to innovate may not be due to lethargy; subsistence farmers can take only limited risks and make minimal investment of money and labour. Investment of effort or money must be seen to give quick and significant returns. Farmers may well regard low yields with security as preferable to higher yields with similar or increased risk (Hudson, 1987, p9).
- Runoff agriculture or SWC may be one part of a multi-faceted livelihood; they must fit in and compliment other activities (Odemerho and Avwunudiogba, 1993).
- Smallholders often farm less than a hectare and can be very reluctant to adopt any measures (such as terraces) which they see as taking land out of production.
- Community woodlots may seem a wise innovation to developers, but local people may suspect them of harbouring pests and of damaging nearby crops. These issues must be checked and recipients may have to be educated and involved to ensure support.
- Risk of improving lower altitude landuse, but of neglecting problems at higher altitude which can damage the former (eg valleyside terraces are improved, only to have landslides change them because slopes above were overgrazed). Solution may be a watershed approach.
- Innovation, like the green revolution in the 1960s and 1970s, may not be scale neutral – if one group benefits, some individuals will get richer and others will suffer or may violently object.

What helps innovation or upgrading to succeed?

Those promoting SWC and runoff agriculture must be aware of the barriers to adoption in a given locality (of a developed or developing country); this requires study of landusers' perceptions as well as capabilities and constraints (Smit and Smithers, 1992; Lutz et al, 1994a). It may be difficult to assess why yields have increased after SWC or runoff agriculture were introduced: is it increased moisture retention; is it trapping of more soil and organic debris to improve fertility; is it better care of land by farmers; is it increased fertilizer or improved seed use? Identifying the key improvements and why something works can be challenging (Murray, 1979; FAO, 1991). Realistically, SWC or runoff agriculture have to be constructed and maintained by users; they must therefore be appropriate. Low-cost and minimal external inputs are desirable for assisting smallholders to upgrade or rehabilitate their landuse (Gischler and Fernandes Jauregui, 1984; Ellis-Jones and Simms, 1995).

Improved yield and sustainable production by intensifying agriculture requires inputs of labour and possibly agrochemicals. If land is plentiful, culti-vators may be better rewarded, or must perceive it to be the case, by extending the area under agriculture using less intensive methods. Where population is low and land relatively plentiful, there will probably be little division of labour and few specialists. Investment of labour or anything else under these condi-tions will yield limited returns, even if there is a good crop, because of poor communications, low-value produce and storage difficulties. When population increases, agriculturalists have limited choices if they are to survive. Options include:

- extend the area under agriculture (assuming there is enough suitable land available);
- intensify;
- migrate elsewhere;
- carry on as if nothing has happened and suffer reduced or failed harvest in due course.

If aid is offered a risk is that it will be accepted and then, instead of the second option, the agriculturalist will pursue the others.

A number of researchers have looked at the transformation of agriculture from extensive to intensive (the second option) the Boserüpian view is that if population increase is not overwhelmingly rapid, agriculturalists can make technological innovations and invest labour (Boserüp, 1965; Conelly, 1994; Tiffen, 1995). Tiffen et al (1994) assessed the intensification of agriculture in the Machakos district of Kenya, where various SWC techniques, including terracing, stone lines, vegetative barriers, trash lines and check dams, were adopted by farmers and have led to improved production and reduced environmental degradation (see also Turton and Bottrall, 1997). Their sugges-tion was that the following factors encouraged and supported improvement:

- evolution of land tenure from communal to individual control;
- prior knowledge of indigenous and introduced SWC techniques;
- a tradition of community organization;
- favourable access to markets;
- remittances from migrants which were invested in SWC.

The crucial question is – and it was addressed by Tiffen et al (1994, p275) – how unique or replicable is the Machakos experience. They conclude that it is being replicated and could be even more widely copied.

In Machakos district *fanya juu* terracing spread quickly since 1979, when the Swedish International Development Agency (SIDA) offered support. Self-help groups, often women, constructed an average of 1000 kilometres per year until, by 1992, almost 70 per cent of the district's cropland was terraced (Postel, 1992, p118). This is a traditional African SWC (ditch-and-upslope bund) technique, which is quite labour demanding to construct (by hand), but gives more or less immediate results: crop yield increases of about 50 per cent are the norm, together with considerable reduction of soil erosion and increased security of harvest. The first and last benefits gain farmers' support.

According to Atampugre (1993) the success in spreading the use of stone lines (*diguettes*) in West Africa reflects:

- lessons learned in the Negev and applied to Burkina Faso (Yatanga province);
- promotion of appropriate levelling devices (such as the water tube – see section on levelling methods later in this chapter);
- resistance of stone lines to damage and ease of repair;
- use of local material;
- minimal labour and skill inputs;
- semipermeable lines that reduce risk of progressive failure;
- less labour required than terracing;
- accessibility even to the very poor and women;
- stone lines that do not undermine existing livelihood strategies;
- increased crop yields;
- reduced drought losses, improved security of harvest;
- conservation of soil and improved fertility by trapping organic matter;
- easier planting and sowing;
- crop diversification.

Demonstration farms and dedicated, competent, adequately funded extension staff can be crucial for expanding and improving SWC and runoff agriculture. It may be sensible to encourage some farmers to do a limited amount of improvement to see how it works and to spread labour demands and other investments (if any) over a long period of time. Most smallholders will adopt an improvement if it gives better security and raised yields, but if it involves risks, too high costs, or takes much time for them to reap benefits, there will be less enthusiasm.

Problems arise when common land is to be treated for SWC or runoff agriculture. Where landusers are not land owners, but have traditionally used the land, terracing, tree planting or other upgrading may attract the attention of the owner and cause conflict. The same problems may occur if common land is improved since those who seek to benefit may not be those who have worked on improvements. Under the law of some countries, digging ditches or building terraces allows claim of tenure and this can result in speculation by outsiders or conflict between land claimants. A solution might be to grant those involved in upgrading to SWC or runoff agriculture usufruct rights in return for a good standard of maintenance.

Problems upgrading to terracing

Since it is known to have huge potential, terracing has been widely promoted, sometimes with poor success at improving yields or leading to catastrophic land degradation, or even both. For example, Belsky (1994) examined the negative responses of smallfarmers to bench terracing in upland Indonesia. Poorly laid out and anything less than well constructed and satisfactorily maintained, terracing can suffer progressive failure, causing severe gullying. Damage may be initiated by pathways, roads or abandoned fields and may spread to well-constructed and managed terraced land, so it is important that surrounding land is checked and, if need be, protective measures taken. Where terraces are poorly drained, soil fertility may decline, or soil crusts may form, or the waterlogged soil may become acid. Soil changes must be constantly monitored and human responses to terrace introduction need to be regularly assessed (Lewis, 1992). Before promoting terracing, alternatives should be reviewed; it also makes sense to establish whether local terracing has expanded or contracted in the past, what the status is at present, and what leads to expansion or contraction (for a study of terracing expansion and contraction in Peru, see Guillet, 1987).

For smallholders, terracing may involve real or feared loss of cultivable land and too much burden of maintenance; therefore, even with aid, they may be reluctant or unable to adopt it. Problems are particularly likely when terracing is installed using machinery which farmers have poor access to in the future.

Grants, subsidies and incentives

Incentives have been tried, especially in developed countries, to encourage landusers to adopt or expand improved land husbandry (Hudson et al, 1993). Richer agriculturalists may be able to afford innovations and may be willing to experiment; poorer landusers tend to be reluctant to experiment because they have little to invest and no security, so all their efforts are focused on tried survival strategies. Aid can provide vital support to enable the poor to experiment and innovate with less risk (Wardman and Salas, 1991; Ewell et al, 1993).

Aid may take the form of providing construction materials, tools, extension staff, training, or food for work. For example, the US Agency for International Development (USAID) supported SWC in highland Bolivia: the motives were to promote economic development and to counter narcotics growing. The results were promising and were gained by effectively focusing scarce development funds (Hanrahan and McDowell, 1997). While the development of SWC and runoff agriculture may be crucial for feeding countries, the payback on any investment can be slow and results uncertain. Hanrahan and McDowel (1997) found considerable variation between ten case studies of USAID-supported smallholder SWC for moisture conservaton efforts, which aimed at improving incomes and employment creation. Yield gains from SWC are often limited but it may still have considerable positive impacts (such as soil conservation, fertility maintenance, crop diversification opportunities etc). Improving traditional irrigation systems had better results but benefitted fewer people.

In the steep lands of the Dominican Republic reasonable success followed USAID efforts to reduce erosion and improve smallholder farmers' incomes (Carrasco and Witters, 1993). For Central America and the Caribbean, Lutz et al (1994b) suggest two factors that are especially important in determining farmers' responses to SWC: expected economic payoff from adopting measures; and the extent to which the practices are known by the farmers at the outset. They also found that farmers were more likely to adopt low-cost methods, such as vegetative barriers, rather than terraces and other substantial structures, and that site-specific factors were crucial.

In the US, Canada and the EU, grants and subsidies have been used to try and encourage SWC. Grants, subsidies and incentives are not always a good idea (Grepperud, 1995). Treacy (1989) argued that full subsidies are only needed if landusers cannot see the need for SWC and better land husbandry. Cash or food incentives intended to encourage Peruvian smallfarmers to terrace their land had the effect of slowing the pace of work, increasing dependency and even led to demands from local communities for grants (Treacy, 1989). Aid may undermine people's sense of responsibility and involvement, leading to poor-quality workmanship and less motivation. People may feel that they are 'working for the government' rather than helping themselves. There is also a risk that one group of people may get paid or receive food, while another gets less or none, leading to discontent and poor morale among the latter. Food for work may be of value when there is a need to distribute food aid, but in practice design and management of these schemes has often been poor (see later in this chapter) (Hudson, 1987, p51). If people are given food or cash to build structures they may come to expect government or aid agencies to pay to maintain the structures. It is important that participants do not see themselves as mere labourers or get too dependant (Ewell et al, 1993). Cash or food aid may be justified if there is a severe land degradation problem and if there are off-farm impacts, such as siltation of reservoirs or irrigation channels.

Food-for-work or cash-for-work support for runoff agriculture and soil and water conservation

Food-for-work has been used to support SWC and runoff agriculture in a number of countries. In Ethiopia, during the 1980s, the European Economic Community (EEC), USAID and various NGOs supported large teams of labourers with food-for-work, building over 1.5 million kilometres of bunds and around half a million kilometres of terraces, together with considerable tree planting (Leach and Mearns, 1996, p202). How much of these structures have been maintained after aid agency staff moved on is difficult to establish; people were reported to be less than fully supportive, claiming the terraces and bunds hindered arable culti-vation, harboured rodents and had reduced yields by bringing subsoil to the surface and burying topsoil. Peasant associations had also largely been required to undertake the work. A more participatory approach might have overcome some of this hostility.

Pretty and Shah (1997, p8) review various food- and cash-for-work efforts during the 1980s, noting some serious difficulties. Clearly, recipient involve-ment must be encouraged and education should accompany food- or cash-for-work efforts to try to avoid these problems. Citing kilometres constructed as a measure of success is unwise if, once the aid ceases, mainte-nance fails and there is abandonment. Success will require more bottom-up support, development of necessary local institutions and, above all, support by local people.

In spite of these problems, food- and cash-for-work schemes can quickly generate large amounts of labour for relatively simple tasks. This might be valuable if authorities are seeking to avoid or recover from some disaster or are installing, for example, irrigation canals or water cisterns. These forms of aid are also regarded by some as a means of doing something to avoid social unrest. Some of the New Deal projects in the US during the Depression were partly undertaken for these motives; the same is probably true of some of the agricultural improvements and SWC measures in north-east Brazil (Hall, 1978).

In the Cape Verde Islands since the early 1970 drought, land degradataion and widespread unemployment have been tackled using *frentes de trabalho* – government-organized workgangs building anti-erosion check dams and terraces, using mechanical means and amounts of labour that landusers cannot match during repair or maintenance, mainly on state land in return for wages. While the *frentes* provide a form of poor relief, and have supported the construction of a great deal of SWC structures and the planting of many trees, the approach is not focused on local needs, fails to motivate landusers to maintain structures, and the workforce has been criticized for becoming over institutionalized. While the real value of these measures for SWC is probably limited, there is little alternative employment; therefore, the *frentes de trabalho* will probably need retuning to better involve and motivate landusers (Haagsma and Reij, 1993).

Food- or cash-for-work aid must be carefully managed or it will not initiate sustainable SWC or runoff agriculture and may not help maintain indigenous practices (Reij et al, 1996, p25).

Tools-for-work or village facilities as incentives

Workers on SWC or runoff agriculture schemes may be rewarded with hand tools (such as hoes, axes and spades) or improved seeds, rather than cash or food. Tools can be used for maintaining structures and improving agriculture and may be a means of avoiding some of the dependency that food or cash aid might cause. Another possibility is to offer community facilities in return for group labour. In Niger in the late 1980s labour on several SWC schemes was rewarded by providing the villagers with a new well, hand pump or school room and equipment (Reij et al, 1996, p57). The problem with this is that special interest groups may choose and 'highjack' the facilities, and there is a risk that the facilities will not be maintained because often no one person is responsible and skills or repair materials are unavailable. Rewards must be carefully selected for a given community.

Working with landusers

Efforts to halt the breakdown of existing agriculture or to upgrade or change it should be based on sound knowledge of the current situation, including information on: the state of the resource base, constraints on production, official policies, and indigenous ways. In practice, it is common for those planning or managing development to lack even basic information. There have been efforts to find solutions to the problem of poor baseline data; for example, Langworthy and Finan (1997) tried to assess the state of a region's agriculture (that of Cape Verde) from ethnographic study. There is also a need to appraise the available strategies and techniques to get the best out of working with landusers (Ellis-Jones and Simms, 1995).

For some years there has been growing support for participation, bottom-up or grass-roots approaches and empowerment (Reij, 1988; Kaihura and Mano, 1993; Kovalo and Nehanda, 1993). A participatory approach can also be a way to bridge exogenous and endogenous expertise (Reij et al, 1996, p25). Unfortunately, participation by local people is still often superficial and results have been mixed (Dhar, 1994). There are occasions when local people resist or are unable to successfully undertake innovation for a variety of reasons – for example, they may be too poor or insecure; they may be sceptical of the value of the proposals; they may have a socio-economic, cultural or religious objection; they may have experienced, or been told of, failed attempts to innovate, and so lack trust. As already stressed, a major requirement to motivate most landusers is to convince them that innovation gives worthwhile returns in the form of better yields or security. In an increasingly market-oriented world, anything which offers cash-earning opportunities is likely to be attractive (Huszar and Cochrane, 1990). Producing crops is not enough, innovation will fail if there are inadequate roads, access to transport or difficulties marketing the produce at a fair price.

There are situations when it may be desirable for people to be mobilized for land improvement, biodiversity conservation or anti-environmental degradation efforts. Mobilization can be through encouragement, appealing to people's sense of national duty, or it can be virtual coercion (*levée en mass*). Consequently, mobilizations have mixed results. Even when people act with enthusiasm there is often a lack of follow-up and ongoing maintenance inputs: a village may turn out to plant saplings when a state minister calls, but individuals are less likely to prevent goats from straying or to take care of the growing trees. Mobilized people may have enthusiasm but lack care, terraces or check dams may be poorly built, or tree seedlings may be mishandled and neglected if there is not good coordination. Nevertheless, there are situations, especially when there is an immediate threat, where mobilization is valuable (for instance, to prepare for, or react to, floods; to stabilize moving sand; to find and destroy pest animals or plants) (Rwejuna, 1994).

Good agricultural research and extension is important; research stations, government agencies and funding bodies strive to improve research, and efforts have generated a huge literature (Greenwood, 1986; Haag et al, 1988; Sombatpanit et al, 1997). Some of the extension knowledge base has come from rain-fed agriculture or rangeland development, some from irrigated agriculture, and some is relevant to the promotion of runoff agriculture. Broad points can be gleaned from the literature.

- Agricultural extension should be a two-way process, gathering information and communicating with landusers as well as offering advice and support – this is embodied in the farming systems research and extension approach (FSR&E) and various forms of participatory rural appraisal (PRA) (see Glossary) (Wang, 1984; Haverkort et al, 1991).
- Well-conducted pilot projects should get higher priority.
- Training and demonstration exercises should be improved.
- International agricultural development institutes, universities, and national agricultural agencies should pay more attention to SWC and runoff farming development.

Extension bodies and NGOs must ensure that they are aware of existing community organizations (such as farmer groups and cooperative labour groups) so that any intervention can fit with local needs and capabilities (Batterbury, 1994; Ndiaye and Sofranko, 1994). Propaganda can be a very effective extension method, disseminating information and motivating, and may be achieved through radio or TV light entertainment as well as specialist broadcasts, training films or live-theatre groups.

Supportive institutions and social organizations

Individual farmers (or other landusers) generally need to cooperate with others over water supplies, sharing of labour, and possibly to develop strategies to support each other if things go wrong. Institutions help to formalize and control relationships (Matin and Yoder, 1987). Institutions have been defined as 'complexes of norms and behaviours that persist over time by serving collectively valued purposes' (Uphoff, 1984). Social organizations are likely to be involved in marshalling work parties to extend, repair and maintain more substantial structures such as tanks, cisterns and larger terraces or bunds. A prolific literature has accumulated on social organizations involved in irrigation, much of it concerned with large scale, gravity-fed canal irrigation, some with indigenous, small scale irrigation, much less with SWC and runoff agriculture (see Coward, 1979). It is difficult to upgrade or extend SWC and runoff agriculture if there are no suitable institutions, and where this is the case, it will be necessary to build them (Critchley et al, 1992, p62). Where several landusers share a source of water, a road, equipment or have other common interests, such as safe disposal of excess runoff, some form of association is vital and tends to arise without outside intervention (Stern, 1988, p2).

SUPPORTING RUNOFF AGRICULTURE AND SOIL AND WATER CONSERVATION WITH LAWS AND REGULATIONS

There are laws aimed at discouraging the clearing and cultivation of steep slopes, and laws seeking to prevent rain-fed agriculture where precipitation is inadequate (usually legislation simply prevents rain-fed farming beyond an average rainfall isohyet on a map). Clearly, with appropriate techniques and management, these laws may be an irrelevant hindrance. Some countries have regulations controlling what soils a crop may be grown on or blanket bans on certain 'land damaging' crops. There are regulations supporting the promotion of SWC or runoff agriculture in a number of countries; however, soil conservation services or departments of agriculture mainly seek to educate and encourage landusers to adopt better practices (Zinn, 1988; Holloway and Guy, 1990). The US has one of the longest established soil conservation bodies, dating from President Rooseveldt's creation of the Soil Conservation Service in 1933 to promote soil protection in response to the Dust Bowl disaster. Because soil conservation issues cross state borders, the US Congress passed the Standard States Soil Conservation Districts Law in 1937 to encourage state legislatures to create soil conservation districts. Today, the whole of the US is covered by such districts (Anon, 1996).

In developing countries laws relating to SWC may be based on those of former colonial nations, augmented by custom and practice, or there may be indigenous laws (Caponera, 1973). Water rights and laws associated with traditional irrigation have generated a sizeable literature, but much relates to rotational sharing of water supplies and is not relevant to SWC and runoff agriculture, the practitioners of which often manage their own supplies. Laws relating to stream sharing and diversion and safe disposal of excess runoff are often applied to SWC and runoff agriculture.

THE ROLE OF WOMEN IN RUNOFF AGRICULTURE AND SOIL AND WATER CONSERVATION

It is women who often adopt SWC and runoff agriculture (Khasiani, 1992). Therefore, gender issues must be carefully examined before attempting to uprate or spread SWC or runoff agriculture (Sachs, 1996). There is a growing literature relating to rain-fed agriculture and irrigation, some of which is relevant (a bibliography is presented by Verkruysse, 1992), and to land degradation (Thomas-Slayter and Rocheleau, 1995). Even if menfolk are absent, they may still decide what is grown and how, and in all probability take a large share of any profit. Women may also have problems getting credit, agricultural inputs, and extension service help, and often do not enjoy the same rights to land or water as men do (De Fraiture, 1991; Khasiani, 1992; Zwarteveen, 1997). Where SWC and runoff agriculture demand considerable labour input, it may be necessary to provide assistance or modify techniques – for example, dumper trucks or wheelbarrows.

PAYING FOR RUNOFF AGRICULTURE AND SOIL AND WATER CONSERVATION

Much of the support for poor farmers will have to be aid, rather than the provision of credit because they have difficulty repaying, and in consequence lenders must force slow payback and high rates of defaulting. Countries vary a great deal in their willingness to spend on SWC and to control environmental degradation; for example, Canada spends far more per capita than the UK, yet both have problems with soil degradation. Worldwide, in rich and poor countries the need for SWC and better land husbandry enjoys too low a profile and far too little funding.

There has been considerable debate about the value of taxation strategies for supporting SWC and improved land husbandry (Kottman, 1984; Seitz, 1984). In the US, tax rules now effectively treat any permanent improvement of the land (through erosion control or reduction of pollution) as a capital expenditure allowable against tax (Barker, 1996). Although it has been claimed that while Australians have a weak regard for land husbandry, there has been progress since the 1980s, which can probably be attributed to Australian income tax allowing a 100 per cent write-off of expenditure on structural works for SWC (Chisholm and Dumsday, 1987, p239).

APPROPRIATE TOOLS

It is important that landusers have access to appropriate tools for construction, maintenance and repair of SWC and runoff agriculture structures, and, if practising arable cropping, for tillage. Tractors are not suitable for narrow terraces, steep slopes, and where soils are soft or easily compacted, nor where farmers are too poor to afford hire charges, and where there are inadequate facilities for maintaining and providing spares. Tractor hire schemes and cooperatives, at least in Africa, have had mixed results, a good proportion discouraging.

What are the alternatives? The first is smaller tractors – widely known as micro-tractors or estate tractors, these are familiar in developed countries in market gardens or parks and large private estates. These may be appropriate for farming steep lands along the Rhine or in Alpine Europe but in developing countries offer fewer economies over full-size machines than might be expected. Secondly, mule or other draft animals may be especially useful in steep terrains and where terraces are too small for tractors. The main problem is provision of adequate fodder. If stall fed, the manure can be easily collected and spread on cropland or pasture. The third option is hand-steered tillers rotovators), originally developed for market gardening in developed countries. Versions suitable for smallfarmers are widely used in China and some can also be used to transport people and materials. Hand-steered petrol or diesel-engine tillers have been adopted for use on terraced land in some countries (for instance, in Peru); they are far cheaper to buy and run than micro-tractors, are easier to maintain, and can be turned on much smaller plots or narrow terraces. They demand no fodder, cause little soil compaction and can be moved from plot to plot fairly easily, even where roads are poor.

Other appropriate tools for runoff agriculture and SWC include: wheelbar-rows, levels for establishing the contour (see later in this chapter), improved hand hoes, and occasional provision of carts or trucks to move stones for construction. A number of NGO bodies are active in developing appropriate technology (for instance, the Intermediate Technology Development Group, Rugby, UK) and there are journals devoted to the field (see *Appropriate Technology*). Without access to blacksmiths or mechanics, repair of simple tools and servicing of equipment such as hand-steered tillers is a problem and steps may have to be taken to provide these services.

Tractor-pulled tools are used on larger farms or by development agencies and government bodies for constructing SWC measures. In particular, ploughs can be adapted to form tied ridges, orchard terraces, broad bed and furrow systems and microcatchments; mechanical scrapers can be used for tank excavation and roaded-catchment formation.

APPROPRIATE LEVELLING METHODS

For many SWC and runoff agriculture techniques it is vital that the contours are reasonably accurately established; this can be quite difficult, especially where topography is complex, in gently inclined areas and where land slopes in more than one direction. Professional survey teams are unlikely to be available for farmers who are poor or who live in remote areas; however, a number of appropriate methods have been developed which can be easily mastered by smallholders, or could be used by travelling 'bare-foot surveyor' teams (Layzell, 1968; Collett and Boyd, 1977). One of these is the *water-tube level*,

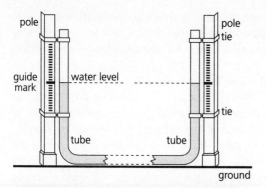

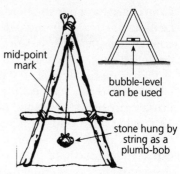

(a) Water-tube level. Based on two poles with graduated scales, to which is attached a length of clear plastic tube (or two clear tubes joined with opaque hose). The device is prepared on a level surface, as in the diagram; next, if the upslope pole is placed on the ground, the water level on the downslope pole rises, showing elevation difference. This can give accurate readings for fixing contours on gentle slopes (<2%) where it is difficult to guess well enough to construct satisfactory SWC measures.

(b) A-frame or levelling triangle. A wooden isosceles triangle is fitted with a plumb line (or a bubble-level on the crossbar); when the triangle is vertical the plumb line hangs at midpoint of the crossbar, and the device can be used to mark out a horizontal plane (with plumb line at mid-point the base of the two feet are at the same level).

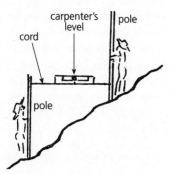

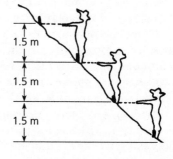

(c) Carpenter's level and cord. Two poles, a cord and a carpenter's bubble level can be used to fix a horizontal plane (bubble is centred when cord is horizontal, so base poles are on same level).

(d) Use of arms to determine vertical distance from a known contour. Once a contour has been fixed by some method this approach can be used to fix others (obviously using the same operative!).

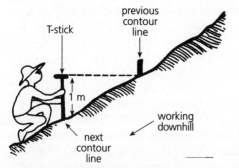

(e) T-stick. May be used to fix other contours from one that has been established.

Note: schematic drawings, not to scale

Source: (a)–(e) drawn by author from a number of sources, including Layzell (1968); Pacey and Cullis (1986, 169); Partap and Watson (1994, 121).

Figure 6.1 Simple devices for determining and duplicating contour lines

constructed from a length of clear plastic tubing (or opaque tubing with clear tubes at the ends). A number of other techniques have been developed, including the *A-frame* and *plumb-bob* method (see Figure 6.1) (Hudson, 1992, pp89–90).

MANAGEMENT OF RUNOFF AGRICULTURE AND SOIL AND WATER CONSERVATION

The *watershed management approach* (or catchment approach) to SWC and runoff agriculture has been promoted by bodies, such as the FAO (Bochet, 1983) and the US Soil Conservation Service, and used in a wide range of countries for agricultural improvement and integrated area development efforts (Singh, 1976; FAO, 1977; Kranz, 1979; Easter et al, 1986; Pereira, 1989, p170; Biswas, 1990; Lopes and Meyer, 1993; Batterbury, 1994; Edwards, 1995; Singh et al, 1995; Thurlow and Juo, 1995). The advantage of a watershed management approach to organizing resource development, runoff agriculture and SWC is that, at least in theory, it can relate to physical structures from top to bottom of the catchment. It is also a useful size, non-ephemeral, and a functional unit for monitoring, research, planning and management. It is possible to coordinate agriculture, groundwater, surface runoff, soil conservation, vegetation and streamflow management, pollution control, and many other useful things within a basin. By adopting a landscape unit for planning and managing SWC and runoff agriculture, those involved are more likely to be sensitive to people, the environment, topography, and whatever else is relevant, rather than just focusing on engineering or economic aspects (Sauer, 1991).

After some 50 years of patchy and unsustained SWC, the Kenyan Ministry of Agriculture adopted a catchment approach in 1988 to concentrate technical effort, aid support, encourage interdisciplinary management and facilitate local participation. The results were encouraging, with rising yields, better SWC, more tree planting, and diversification of the approach (Pretty et al, 1995). This is not an isolated success story; integrated watershed management is seen by many as an effective route to sustainable development and one which will encourage participation of the landusers (Maaren and Dent, 1995).

However, hydrological boundaries do not always correspond with administrative or tribal boundaries. There are other weaknesses. Critchley et al (1992, pp30; 71) noted that while the watershed planning and management approach may have theoretical strengths, practitioners should treat it with caution because landusers may not perceive their surroundings on a watershed level and are more likely to work at the scale of their plot, as well as that of a few neighbours. With these weaknesses in mind, some have claimed that the watershed unit makes sense to managers and administrators but may not suit actual landusers.

Some advocate an individual or small-group scale of management rather than a watershed scale. Reij et al (1988, p77), Critchley (1991), and Critchley et al (1992) suggested it was better to organize through village groups (village land units), or farmer groups, or deal with individuals rather than adopt a watershed approach. Hudson et al (1993, p14) noted that within a watershed there can be considerable diversity of farmers' abilities, attitudes, needs and resources – getting them to effectively cooperate and develop suitable practices may be difficult. Therefore, a unit smaller than a watershed might be better. In practice, the watershed units used are micro-watersheds of roughly 500 to 1500 hectares in area – the catchments of small tributaries, begging the question: what unit is to be used in lower basin, where topography means bigger areas and more farmers? Turton and Bottrall (1997, p3) also urge caution toward the watershed approach, although they accept some regions do have traditions – for example, tank irrigation management by community groups, which might support an effective watershed approach.

In steeper regions a watershed approach may be more desirable; Bochet (1983) felt that in these regions it can support community involvement and cooperation. Advocates of watershed management feel that it promotes collective action and cooperation by groups of land-users to control runoff and counter erosion (White and Runge, 1994; 1995a; 1995b). Furthermore, it has been argued that there are regions where people are too dispersed, or are living in hamlets or scattered farmhouses for a village approach to work (the case in Cape Verde, according to Haagsma and Reij, 1993).

There have been attempts to develop a participatory watershed approach (use of a watershed unit with participatory local-level planning and management) as a way of getting NGO, community organization and government department collaboration, cooperation and coordination (Michaelson, 1991; Farrington and Lobo, 1997). Hinchcliffe et al (1995) outline a participatory approach to watershed development. Whether or not a watershed unit of management is adopted, and even where land is under communal ownership, a cooperative approach is likely to be important (White and Runge, 1994; 1995a; 1995b).

USAID, seeking to achieve sustainable SWC and to improve smallfarm incomes by using a watershed approach, supported a project in the Ocoa Valley, Dominican Republic, in 1982. Carrasco and Witter (1993) report on a 1990 appraisal of those efforts.

The *community management approach* has become popular in recent years. And even if a watershed management approach is adopted, it must adequately consider local institutions if it is to work (Bottrall, 1993). Lopes et al (1993) called for watershed management to better evaluate people's needs and environmental limitations, rather than to focus, as is often the case, just on SWC engineering or forestry. The complexity of issues that must be dealt with even within small watersheds was stressed by Thapa and Weber (1995). Pretty et al (1995) saw potential in Kenya for mobilizing communities and getting an interdisciplinary approach, by using catchment committees to link various agencies and promote local needs.

Today, many support a community approach to SWC, improved land husbandry, and runoff agriculture, although some, like Wardman and Salas (1991), argue that it is best to adopt an individual landuser approach which demands little cooperation and therefore has less chance of complications and failure. One problem is the extension of promising approaches from research station or pilot project to large numbers of farmers. Sometimes this takes off and techniques spread from landuser to landuser; elsewhere NGOs or governments may need to trigger or support the spread, probably focusing on selected subgroups, such as women or progressive farmers.

Pretty and Shah (1997, p11) list several highlights of community-based SWC and land husbandry improvement, including successes in Australia, with one third of all farmers involved in voluntary community schemes that help them deal with environmental problems. Pretty and Shah (1997, p14) stress the need to build up and strengthen local institutions to support SWC. Not only must farmers be motivated, but communities as a whole. It is important that before any efforts are made to upgrade or spread SWC or runoff agriculture, the developer checks whether landuse and land ownership laws and rights, and water laws and rights, will be supportive. Sustainability and good land husbandry are as important as yield improvement (Sombatpanit et al, 1997).

Community organizations have potential as bodies for managing SWC and runoff agriculture (Batterbury, 1994). There are traditions of this sort of approach in Nepal, where elected village councils (*panchayets*) have acted as traditional management units for the upper parts of watersheds. In the People's Republic of China local communities have often been able to construct and manage SWC and runoff agriculture projects. The important thing is that there is no hindrance from local feuds, caste, tribalism or political factions, as is often the case.

A LONGER-TERM VIEW IS NEEDED

Agriculturalists usually know an environment very well and can warn of risks and assess the potential of proposals in a way that exogenous experts may be unable to match. Participation of local people from the start of the planning process can therefore bring planning benefits; this also helps people to welcome and support innovations (Kidme and Stocking, 1995). Too much development is based on a project-by-project approach and adopts an inflexible fixed-programme approach. There is a need for flexibility, a longer-term strategy and an overall policy.

Management measures and community cooperation can be more effective than physical measures and inflexible legislation. For example, careful monitoring and agreed control of grazing, or a shift from grazing herds to policed fodder collection and stall feeding, can avoid vegetation and soil damage – without any digging of bunds.

7   THE FUTURE FOR RUNOFF AGRICULTURE

An objective attempt to assess the future for runoff agriculture should consider: the challenges it must face; the demands and needs it should address; possible advances in techniques and strategies; and whether there are complimentary, supportive or better alternatives.

THE CHALLENGES RUNOFF AGRICULTURE MUST FACE AND THE DEMANDS AND NEEDS IT SHOULD ADDRESS

There have been attempts to predict the future for runoff agriculture and SWC (see Jeffords, 1987; Anon, 1996a; special issue: 'Towards the Next 50 years', *Journal of Soil & Water Conservation,* vol 51, no 5, especially pp444–448). Demand for water is clearly rising worldwide as irrigation spreads, populations increase, industry expands, sewerage is installed, and as the process of development prompts greater per capita usage (Crosson, 1995). At the local and regional level there is considerable variation in demand for domestic water supplies, reflecting per capita incomes and ability of administration to provide piped water. Industrial demand will grow in certain areas and many cities are expanding enough to pose regional or even national water supply problems. In some countries, urban and rural water demands will compete, as is already the case, for example, in parts of southern US and Europe. Even in temperate and humid environments there will be rising water demand in lowland areas to supply large-scale irrigated agriculture (Bartolina, 1996).

It is on higher ground (mid and upper catchments), and in remote areas, that runoff agriculture will prove important, supporting landusers by supplying moisture where there may be little other way, and – by reducing erosion and stream silt loads – helping to stabilize streamflow and improving groundwater recharge. By doing all this it will also play a vital role in maintaining supplies to, and reducing siltation of, lower catchment water users, including large scale irrigation.

At present governments tend to encourage and support lowland commercial agriculture, the produce of which can depress domestic agricultural prices, reducing the profits of smallholders in uplands who then are forced into land abandonment or misuse. This leads to water quality and quantity difficulties that affect lowland agriculture. Therefore, neglect of one sector can have an unwanted feedback on sectors that are supported (including impacts on city and industry supplies). It is possible that mainstream irrigation could be taxed or encouraged to support runoff agriculture higher in watersheds. Runoff agriculture may boost supplies of food and commodities in return for relatively low investment (compared with large-scale irrigation).

There is a need to improve upon the yield, security of harvest and sustainability of rain-fed agriculture, which avoids resort to often unsuccessful mainstream irrigation (Pereira et al, 1996). Runoff agriculture is one of the best routes to more secure and improved livelihoods and more sustainable development, at least for the 'millions of smallholders in the tropics farming under rainfed conditions in diverse and risk-prone environments. In a constant struggle to survive...' (Reijntjes et al, 1992, pxvi). Reijntjes et al (1992, p2), tried to assess the qualities needed for low external-input sustainable agriculture which would cut environmental degradation and improve rural livelihoods, and concluded that such agriculture should be:

- ecologically sound;
- economically viable;
- socially just;
- humane to people and livestock;
- adaptable.

Runoff agriculture and SWC techniques can meet most, if not all, of these demands. Per capita, the world's food production is slowing; efforts need to be made to boost food and woodfuel production, especially from the small-farmer sector. Runoff agriculture and SWC could be the way to counter the decline of food and fuelwood availability.

The challenges that must be faced by runoff cultivation include: the possible impacts of global warming; acid deposition/acidification; other forms of transboundary pollution; and socio-economic difficulties. There may be regions where drought is more likely or more intense because of natural climate change and global warming due to anthropogenic causes. Some studies of records and environmental evidence suggest that sub-Saharan Africa has shown a drying trend (Hulme, 1992; Alvi, 1994). Whether as a result of land degradation, global warming, or through natural causes is less clear.

Attempts to model likely future global warming scenarios have so far given uncertain predictions (Parry, 1990). Global circulation models today attract considerable criticism and debate, and precise, reliable predictions are not yet possible, although there have been attempts to identify likely scenarios (Glantz, 1992). It is possible to approximately map the areas that are likely to become drier and which might benefit from development of runoff agriculture and SWC (see Figure 7.1). Better computers and improved knowledge of, for example, El Niño – Southern Oscillation (ENSO) should permit more accurate predictions of droughts and above average rainfalls. It has been suggested that monsoon-type rainfall patterns will alter. Furthermore, evapotranspiration

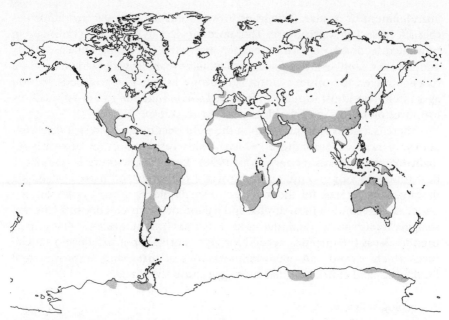

Source: Jones, 1997, p338, Figure 11.8.b. Based on Hadley Centre Global Environmental Circulation Model 1995 data.

Figure 7.1 Areas of the world predicted to get drier with global warming.

rates will increase as a consequence of warming, of altered wind patterns, of changed cloud cover and, regardless of altered local precipitation or evaporative conditions, of photosynthetic variation in response to more carbon dioxide (CO_2) in the atmosphere – as CO_2 levels rise plants may increase transpiration and lose more water.

Conversely, some areas may receive more useful precipitation than now; indeed some may have too much. Ramírez and Finnerty (1996) examined various possible climate change scenarios associated with anthropogenic global warming and increased atmospheric carbon dioxide levels. Their conclusion was that, at least for some parts of the world, the future may not necessarily mean increased demand for irrigation. This is because increased atmospheric carbon dioxide may speed plant growth and shorten growing seasons or alter crop metabolism.

There is a possibility that some mainstream irrigation development may release unacceptable amounts of greenhouse gases, notably methane. This threat could encourage more interest in agriculture which involves less frequent water application and flooding. The problem of nitrification of groundwater and pollution of rivers, lakes, the sea and groundwater with agrochemical-contaminated return flows should help shift interest and investment from large scale, high chemical-input, mainstream runoff agriculture.

Acidification may be more predictable since it can be monitored, is mainly restricted to regions down wind of industrial or urban areas, and affects some vegetation, crops and soils more than others. Where acid deposition becomes a problem, agriculture may have to cope until pollution controls can be implemented. Runoff agriculture should be more adaptable than large-scale irrigated agriculture.

Transboundary pollution is increasing; the 'victim' may not be aware of the problem, or may recognize it but fail to obtain controls in the emitter countries or compensation payments from them. Regional agricultural planners and international bodies must monitor for potential problems. All forms of improved agriculture will need to be vigilant against pest or disease organisms which destroy or debilitate crops or livestock as well as chemical pollutants. Another challenge is posed by reduced stratospheric ozone shielding, which results in UV damage to crops, livestock, infrastructure, people and wildlife. Runoff agriculture is as vulnerable to acid deposition and other transboundary pollution as any other form of crop production, and may be more vulnerable to UV damage than is mainstream irrigation because it is often practised at higher altitudes where UV damage may be greater. It may, therefore, make sense to research more UV-tolerant crops for highland runoff agriculture.

Where agriculture has come to depend upon agrochemicals, there may be insidious chemical-timebomb problems, where pollution steadily accumulates with little obvious sign until a crucial threshold is reached and a serious problem suddenly appears (for instance, acidification may trigger toxic aluminium releases from a high-aluminium content soil; or pesticides may build up and suddenly be liberated from the clays holding them if the climate warms). Runoff agriculture strategies often maintain fertility without the need for agrochemicals, and so may be a way of reducing the incidence of these problems.

Agriculture can be badly disrupted by warfare or civil unrest. In Africa, conflict has been a major factor leading to failed agriculture and land degradation; attempts to promote better landuse must be accompanied by more stable social conditions. A range of factors may drive rural people to migrate to urban areas, seeking employment – a process accelerated if there is decline of profit or risk in agriculture. There are parts of the world where traditional runoff agriculture is in decline because labour is attracted away by mining, employment in the petrochemicals industry or tourism (for example, the Maghreb countries and Yemen). The abandonment of indigenous agricultural strategies means that there will probably be nothing for these people to return to if there is a contraction in employment or when they get too old to find work (Varisco, 1991). Few urban migrants find satisfactory employment, and many face poverty and become a burden to their country.

Once living in a slum, migrants often find it impossible to procure adequate employment and cannot return to their former rural life. Runoff agriculture might help by offering a sufficient livelihood to keep potential migrants on the land. It might also be a means for slum dwellers to find food and improve their livelihoods when little else is on offer. By 1996 an estimated 15 per cent of the world's food was produced by urban agriculture (or peri-urban agriculture), with one in every three city dwellers involved in horticulture (Anon, 1996b). Often governments have done little to help and have even discouraged such agriculture. However, this has begun to change – for example, in Cuba, Zimbabwe, parts of South Africa and India. Runoff use could prove a valuable way of intensifying urban agriculture and might offer a way of coping with debris-charged water from storm drains. The main challenges to using urban runoff are that flows can be sudden and intense, and sewage and other effluents often contaminate flows, posing a health risk to agriculturalists and those consuming food crops. Therefore, it is important to have appropriate effluent treatment (that can cope with peak flows), and careful choice of crops and cultivation methods to reduce these risks.

The key responses to these threats should be better planning and management, which are flexible and adaptive, and infrastructure that can cope with or adapt easily to change. Runoff agriculture faces such a variety of challenges that some have argued that to respond effectively managers must adopt a holistic approach (Biswas and El-Habr, 1993). Existing planning tools such as environmental and social impact assessment could be used more to assess the possible effects of proposed SWC and runoff agriculture developments, so that unwanted impacts can be eliminated and responses to change, as far as possible, incuded (FAO, 1996). Potential sites for runoff agriculture could be better determined using techniques such as GIS (Tauer and Humborg, 1992).

Many developing country governments prefer to invest in large scale production of export crops, and provide much less support, if any, for subsistence farming and food crop production. This is one reason why a number of countries, once self sufficient, now import food and, at least in part, pay for it with industrial development, tourism industry, or mainstream irrigation-produced export crops (obtained by mining natural resources and polluting the environment), rather than sustaining development. There needs to be more interest in promoting the sustainable development of agriculture, especially by small scale producers (which means the added benefit of more jobs created) (Vincent, 1990a; 1990b). A number of researchers have stressed the need for reducing agricultural dependency on external inputs, using what a locality has to offer (Reijntes et al, 1992). A local focus also offers the possibility of better adapting strategies to environment and socio-economic conditions, which many advocate as a promising route to sustainability ('think globally, act locally').

Today, large scale irrigation is an important route to intensifying agriculture, one that can give impressive yield improvements. But it is sometimes environmentally and socially damaging, leads to dependency, may be unsustainable if poorly planned, and its future expansion is limited by a decline in the availability of good soil, suitable water supplies and very high costs. Runoff agriculture and SWC can be local in scale, and a route to intensification with fewer unwanted impacts than many forms of irrigation have. Runoff agriculture can also be a strategy which requires few or no material inputs from outside the locality – something that is attractive to supporters of appropriate and sustainable development and a way of avoiding dependency.

Large-scale, long-distance transfer of irrigation water may seem attractive in countries like the US or CIS, and huge sums have been invested. There are, however, problems in addition to cost and environmental impacts. For example, proposals for piping water from Turkish rivers to the Gulf states, though probably quite feasible in terms of engineering and funding, are almost certain to be rejected because the project would mean dependency and be vulnerable to terrorism or war.

BETTER RUNOFF AGRICULTURE AND SOIL AND WATER CONSERVATION MANAGEMENT

Developing techniques is not enough; there is a need to develop effective regional management plans, rather than rely on ad hoc efforts. Such management should:

- determine priorities;
- prepare contingency plans to deal with problems;
- constantly assess progress;
- be adaptive;
- share information.

It would make sense to: promote a diversity of crops; encourage a participatory approach to research, extension and management; stress sustainable development goals; minimize dependency on inputs from outside the region where runoff agriculture is being developed; coordinate agriculture, land husbandry, conservation of soil and biota; and pay attention to socio-economic needs and try to get integrated, comprehensive resource management. Integrated water resource management is often called for but still tends to give insufficient attention to runoff – a recent text on applied hydrology discusses streamflow, groundwater and even desalination, but not runoff supplies!

POSSIBLE ADVANCES IN TECHNIQUES AND STRATEGY

In the future there will probably be better techniques and wider application of long-range weather forecasting. This would be especially valuable for runoff agriculture if it could show likely usable precipitation well in advance and give an idea of the timing of the onset and end of rainy seasons (Omotosho, 1993; Hulme, 1994) or floodwater flows. Drought watch and early warning is likely to improve with the adoption of remote sensing and automatic instruments in the field, and computer processing of data (Kogan and Sullivan, 1993). There has been considerable interest in recent years in the value of ENSO events as a means of long-range forecasting (12 to 24 months' warning) of drought in Africa, EurAsia and Austral–Asia. El Niño events mean increased frequency and intensity of downpours in western South America and the western US, which increases the chance of damage to structures such as terraces, check dams and reservoirs. Monitoring El Niño events and similar phenomena elsewhere therefore deserves attention (at the time of writing El Niño changes off South America's west coast were prompting expectations of extreme weather events later in the year as far away as Austral–Asia. Much could be done to make agriculture more secure against these periodic extremes. Any runoff agriculture, SWC or irrigation development in areas affected by El Niño-type events should be able to cope adequately; often infrastructure is damaged because it has not been properly designed.

MODERN MATERIALS

A strength of runoff agriculture is that it can often be implemented using locally available materials which makes it widely accessible to smallfarmers. Modern materials may still play a useful role, provided they are affordable and do not lock producers into a dependency situation. Plastic pipes, channels and some soil amendments may prove useful if they are cheap, robust and easy to transport. This is already the case with some channelling, plastic sheeting, and pipes made from recycled plastic. Another promising range of materials are the geo-textiles; these are fibre mats or cloth designed to stabilize slopes or trap silt. A broad range of these products is already available and widely used by construction and landscaping specialists.

CONTRACT AGRICULTURE

Where runoff cultivators operate on a small scale there may be opportunities to penetrate commercial markets if they contract to a marketing body. This approach has been promoted by the Bodyshop chain as a way for small scale producers to avoid exploitative middlemen, lack of economies of scale and difficulty managing standards, production schedules, transport and marketing. A number of supermarkets in developed countries have been marketing produce from developing country smallholders or from larger farms which use cheap local labour (Adams, 1992, p185; Porter and Phillips-Howard, 1997). The trade, so far, has been in airfreighting high-value vegetables, fruit or cut flowers to Europe and the US, especially during the winter. There has been quite a rapid expansion of these arrangements, which some hail as mutually beneficial and others condemn as exploitation of the poor and their environment.

DIVERSIFICATION OF CROPS

Mainstream agricultural development has concentrated on strategies which produce one or a few crops; a shift towards more diversity, and ideally an integration of crops and livestock, may have advantages for smallfarmers and assist them to achieve sustainable development. Where rainfall is a constraint on diversification, SWC and runoff agriculture may help.

BIOTECHNOLOGY

There is growing interest in biotechnology as a route to more sustainable, environmentally and socio-economically appropriate agriculture (Elliott and Wildung, 1992; Kirschenmann, 1992). Already a number of promising new tools exist; the challenge is now to assemble packages of tools, and management approaches to suit given situations. Biotechnology has already helped and could further aid runoff agriculture in many ways – for example by:

- providing crops with shorter growing seasons;
- providing drought- or salt-tolerant crops;
- improving pest, weed and disease resistence without recourse to polluting pesticides and herbicides;
- providing crops that are capable of biological nitrogen fixation (in effect, they produce their own fertilizer);
- providing improved sources of mulch or compost;
- providing new crops more quickly;
- providing ways of rehabilitating damaged soils (bioremediation).

However, biotechnology is new and poses unknown threats (real or imagined); it is potentially a double-edged sword which could do much, but which might also cause great harm. Any application of biotechnology, to runoff agriculture or whatever, must be subject to very careful precautionary studies (such as impact assessments and trials) before implementation. With commerce likely to be heavily involved, great care and strong supervision will be needed to see that such precautions are adequately taken.

CONCLUDING COMMENT

The value of runoff agriculture and SWC has been widely demonstrated, particularly as something to assist smallholders in improving their livelihoods and in raising food production where there is little in the way of funding. The techniques can also be adopted by larger producers, and offer commercial agricultural opportunities where there are insufficient water supplies for conventional irrigation development. Wherever runoff agriculture and SWC are used there are opportunities for reducing soil degradation and other environmental impacts, which are often associated with improving agriculture. There is, however, a problem – by requiring little in the way of outside inputs these techniques are accessible and reduce dependency. However, this also means that there is little profit to be made by commercial bodies. Therefore, companies are likely to continue to invest in large scale irrigation and not runoff agriculture and SWC, so support must largely come from governments, NGOs and international agencies, and innovations will have to attract smallholder users and be adoptable if they are to spread and be maintained.

GLOSSARY

Local words are italicized and explained in the text at first occurrence. A full list of Acronyms and Abbreviations can be found on page xiii.

agrochemical   Chemicals: fertilizer, pesticide, herbicide, fungicide, etc used for agricultural purposes.

agroforestry   Also termed **agrosilviculture** – strategies which grow annual crops and trees or perennial crops in a way that is mutually supportive, combats soil degradation and makes good use of moisture. **Agropastoralism (silvipastoralism)** has similar goals but substitutes fodder for food or commodity annual crops.

arroyo   (see also *wadi*) Seasonal or periodically flowing stream, for much of the time a dry valley or one with much smaller flow than at the time of the flood.

bund   Earthen embankment, usually quite low (Persian origin?).

catchment   Area contributing runoff to a water course (UK). River basin is more or less synonymous. (see also **watershed**.)

command area   (see **gravity irrigation**.)

dependency   Situation where people or country depend on an outside source (such as aid or loans) for some need. May lead to restrictions on freedom and loss of initiative.

divide   Boundary between drainage systems (watersheds) (US) (see also **watershed** – UK).

El Niño   Periodic weakening of trade winds reduces the cold current along the Pacific Coast of South America; the warming of the eastern tropical Pacific around Christmas is known as El Niño. Linked with the Southern Oscillation (and abbreviated to ENSO), these changes affect climate over a large swathe of the globe, and are manifested as storms and rains over Central America and down the Pacific Coast and, later, as droughts over Brazil and probably in Austral–Asia and Africa. It appears as if it may be possible to predict broad climate changes a year or more ahead once an ENSO event is underway.

empowerment   The improvement of people's ability to secure their own survival and development, and the capacity to participate in and exercise control over crucial decisions affecting their well-being.

erodability   Vulnerability to **erosion.**

erosion   Soil erosion is the detachment and removal of soil material from the surface of the ground, mainly by water or wind (removal might also occur, for example, as a result of harvesting crops with adhering mud).

erosivity   Capacity to suffer **erosion.**

expansion   Expanding agriculture into new land, if it is available. Opposite approach is **intensification.**

expert system   Computer program which stores a body of knowledge and with it helps a user to perform tasks that usually demand input from a human expert.

extension   (1)   Process of aiding and supporting agriculturalists to function – provision of research support, training, inputs, advice. Undertaken by agricultural extension service or body.

   (2)   Seeking to improve agricultural output by opening new land (see **intensification**).

field capacity   Maximum amount of moisture which can be held in a given soil.

FSR   Farming systems research – since the 1970s FSR has gained support as an operational approach to agricultural research and development. It is the integrated use of farm surveys, diagnostic studies and adaptive research. Carried out by a multidisciplinary team with a bottom-up approach, researchers listen to farmers, their families and contacts to understand whole livelihood systems and to identify potential, needs and constraints or risks. Often linked to **extension** efforts and abbreviated to **FSR&E.**

gabbion   Wire-net construction box (typically two by one by one metres in size) which can be filled on-site with pebbles or cobbles to provide a cheap, robust, transportable semipermeable construction.

gravity irrigation   **Irrigation** fed by canal or pipeline without the need for pumping to lift the water; the supply system feeds from a river or reservoir to crops at a lower altitude so that gravity moves the water. The area that can be fed with water by such a scheme is known as the **command area.**

growing season   Period from seed planting (root planting or end of dormancy for perennial crops) until harvest. Period during which crop requires moisture.

headworks   Structures which are designed to extract water from a river, reservoir or lake. May need to be designed to resist flood damage, to prevent shift of stream channel, to avoid

	becoming choked with silt, and to provide a controlled flow of not too silty water.
holistic	Precise meaning can be debated – broadly, an approach that considers all components and aspects of a system. Contrasts to reductionist approach, where components of a situation are dealt with individually in isolation. Holistic approach is said to be sensitive to situations where 'the whole is greater than the sum of the parts'.
indigenous knowledge	Knowledge held by a group of people in a given locality, which may be the sum of many generations of experience – in other words, more than local knowledge because it may incorporate lessons learned or acquired in the past.
intensification	Process of improving agricultural output by getting more from land under use, by increasing inputs of labour, agrochemicals or compost, and better crops or techniques (or a mix of more than one of these). Opposite approach is **expansion.**
irrigation	Any process other than natural precipitation that provides moisture for crops, pasture, tree crops, etc. May provide occasional supplies to boost yields or to prevent crop or pasture loss; may provide whole moisture needs of agriculture. (See also **gravity irrigation**.)
low cost	Within the means of smallfarmers or rural folk.
land degradation	Loss of productive capacity or biota, generally characterized by impoverishment of vegetation cover and consequent soil damage (no universal, precise definition).
land husbandry	Implementation and management of preferred systems of landuse in ways such that there will be no loss of the land's stability, productivity or usefulness for the chosen purpose(s)
marginalization	Process which results in people moving to less than optimum locations or engaging in less than optimum livelihoods. May be caused by many things (civil unrest, persecution, economic forces, lack of opportunity) can happen to people settled in a locality if some factor affecting their livelihood alters (decline in demand for a product they produce, degeneration of communications, climate change, soil **erosion**). May be reversed by improvement in factors affecting livelihood.
NGO	Non-governmental organization – typically a non-profit, voluntary group engaged in relief or development or environmental protection activities.
nitrogen fixation	Ability to convert atmospheric nitrogen into a form that can be used by higher plants (crops) – 'free fertilizer' (known as fixation). May be done by bacteria, algae or other micro-organisms in the soil, in irrigation water or in association (symbiosis) with plant roots).

peak discharge	Maximum flow from a stream or greatest flow of runoff. Channels, storage tanks, etc must withstand this.
PRA	Participatory rural appraisal – similar to **FSR**, an approach which seeks to involve the subjects of research and development in order to benefit from their knowledge and to avoid their alienation. It is a multidisciplinary, bottom-up approach which can empower those involved in development. **RRA** (rapid rural appraisal) is similar, with the emphasis on speed rather than on participation (in practice both are merging). Subjects of development can suggest research and development priorities.
relay cropping	Growing more than one crop simultaneously for part of the growing seasons – for example, beans planted after maize has become established.
return flow	Water released from an irrigation scheme (or a pisciculture or aquaculture facility) – usually contaminated with salts, chemicals, organic waste or disease organisms. On return to streams, water bodies or groundwater, return flow is likely to cause problems (off-site).
RSLE	Revised soil-loss equation (or **RUSLE**: revised universal soil-loss equation). Versions of **USLE** designed to 'fit' conditions other than those for which USLE was developed (continental temperate US); usually 'tropicalized' versions of USLE.
runoff	That part of precipitation which is neither absorbed into the ground, stored on the surface, nor evaporated, but which flows over the land (technically surface runoff). Runoff generation: for a catchment of twice the area of a cropped plot which receives 270 millimetres per year of precipitation of which 30 per cent becomes runoff: $(30/100 \times 2 \times 270) + 270 = 432$ millimetres per year.
salinization	Build up of salts (or alkali compounds: sodification) in soil, surface waters or groundwater. May proceed to point where soil or water is rendered unproductive and can be difficult and costly to remedy (see also **sodification**).
saturation	Condition in which pore spaces of a soil are full of moisture, excluding air (if persistant, soil becomes waterlogged).
semiarid	Environment where there is enough moisture stress to slow or halt vegetative production seasonally or periodically. Crops can grow and livestock can be kept, but there are periods of stress. Precipitation may be 1000 millimetres per year or more but is uncertain and variable. These areas may have quite dense human and livestock populations.
shifting agriculture	Also termed shifting cultivation, bush-fallow, swidden, non-sedentary agriculture, slash and burn, and many other

local names. Soil fertility is maintained by moving the culti-vated plot, rather than crop rotation and addition of inputs. A plot is cropped for one or two harvests, abandoned to recover naturally, and a new one is cleared from natural vegetation. Demands enough land to allow recovery of land before reuse (since recovery may take over 20 years, a farmer needs at least 20 times the land as his cropped plot to sustain production). When popula-tions grow the system tends to break down. Often seen as backward but in reality has many techniques to offer modern agriculture.

smallfarmer   **Small-scale** – these agriculturalists include a large number who practice **subsistence agriculture**: who produce for their own consumption, with little or no surplus for sale. Typically a family unit, using family labour and producing enough for family needs.

smallholder   Virtually synonymous with **smallfarmer**, could include small scale herders and horticulturalists. Essentially a family holding providing enough for subsistence, perhaps with occasional small surpluses of crops for sale (unlikely to exceed 50 hectares, usually far less).

small-scale   Landholdings or tenancies of 2 to 20 hectares (exception-ally up to 100 hectares).

sodification   Sodification (alkalinization) – build up of soda-salts (see also **salinization**).

soil amendment   (1)   Use of chemicals, physical treatments or biotech-nology to rid soil of an unwanted compound. For example, use of gypsum to help rain leach salts from a salinized soil; or use of bioremediation (encouragement of bacterial activity) to release or convert contamination such as waste oil from a soil.

(2)   Addition of compounds to soil to help it hold moisture.

subsistence   All or most of what is produced is consumed by the
agriculture   producer or the producer's family, perhaps with a little sold for cash.

SWC   Soil and water conservation.

tank   Man-made reservoir, designed to catch and hold runoff or floodwater for dry season use. Of simple earth trench construction in most cases. Widespread in India and Sri Lanka, where there is a long tradition.

USLE   Universal soil-loss equation (see text discussion) (see also **RSLE**).

USSCS   US Soil Conservation Service – governmental agency which undertakes soil conservation work (including surveys). Initially, part of its role was to provide employ-ment during the 1930s Depression. Provided a foundation for later US soil conservation programmes and stimulated interest elsewhere in the world. Some rivalry with earlier established US Department of Agriculture (USDA).

vertisol Black cotton soil – soils which tend to swell when moist so precipitation fails to infiltrate. Unless infiltration can be improved, vertisols provide erratic and poor crops.

wadi (see also *arroyo*) – seasonal or periodically flowing stream, for much of the time a dry valley or much smaller stream than at time of maximum flow.

watershed Area contributing to a watercourse (US and more or less worldwide, although UK has in past varied) (see also **catchment**). UK – equivalent of **divide**: boundary between drainage basins.

water-table Upper surface of groundwater.

REFERENCES

Adams, W M (1985a) 'River control in West Africa', in A T Grove (ed) *The Niger and Its Neighbours: environmental history and hydrobiology, human use and health hazards of the major West African rivers.* A A Balkema, Rotterdam, 177–228

Adams, W M (1989) 'Definition and development in African indigenous irrigation', *Azania* XXIV(1), 21–27

Adams, W M (1990) 'How beautiful is small? Scale, control and success in Kenyan irrigation', *World Development* 18(10), 1309–1323

Adams, W M (1992) *Wasting the Rain: rivers, people and planning in Africa.* Earthscan, London

Adams, W M (1996) 'Irrigation, erosion and famine: visions of environmental change in Marakwet, Kenya', in M Leach and R Mearns (eds) *The Lie of the Land: challenging received wisdom on the African environment.* James Currey, London, pp 155–168

Adams, W M and Carter, R C (1987) 'Small-scale irrigation in sub-Saharan Africa', *Progress in Physical Geography* 11(1), 1–27

Adato, M and Miller, J (1986) *Runoff Agriculture: farming arid and semi-arid lands: a review of the integrated research program at the Desert Runoff Research Farms in the Negev Desert of Israel.* Jacob Blaustein Institute for Desert Research, Ben Gurion University of the Negev, Sede Boquer Campus

Agarwal, A and Narain, S (1997) 'Dying wisdom: the decline and revival of traditional water harvesting systems in India', *The Ecologist* 27(3), 112–116

Agarwal, A and Narain, S (eds) (1997a) *Dying wisdom: rise, fall and potential of India's traditional water harvesting systems.* Centre for Science and Environment, New Delhi

Agnew, C and Anderson, E (1992) *Water Resources in the Arid Realm.* Routledge, London

Agnew, C T, Anderson, E, Lancaster, W and Lancaster, F (1995) 'Mahafir: a water harvesting system in the eastern Jordan (Badia) Desert', *GeoJournal* 37(1), 69–80

Agricultural History (1985) Special Issue. Symposium on the History of Soil and Water Conservation, 59(2), 103–372

Ait Hamza, M (1996) 'The mastery of water: SWC practices in the Atlas Mountains of Morocco', in C Reij, I Scoones and C Toulmin (eds) *Sustaining the Soil: indigenous soil and water conservation in Africa.* Earthscan, London, pp 44–47

Alconada, M, Ansin, O E, Lavado, R S, Deregibus, V A, Rubio, G and Gutierrez Boem, F H (1993) 'Effect of retention of run-off water and grazing on soil and on vegetation of a temperate humid grassland', *Agricultural Water Management* 23(3), 233–246

Alegre, J C and Rao, M R (1996) 'Soil and water conservation by contour hedging in the humid tropics of Peru', *Agricultural Ecosystems & Environment* 57(1), 17–25

Alemayehu, M (1992) *The Effect of Traditional Ditches on Soil Erosion and Production* (Soil Conservation Research Report no 22). Centre for Development and Environment, Institute of Geography, University of Berne, Berne

Allen, R R and Fenster, C R (1986) 'Stubble-mulch equipment for soil and water conservation in the Great Plains', *Journal of Soil & Water Conservation* 41(1), 11–16

Allen, W (1965) *The African Husbandman*. Oliver and Boyd, Edinburgh

Altieri, M A (1995) *Agroecology: the science of sustainable agriculture* (2nd edn). Westview Press, Boulder

Alvi, S H (1994) 'Climate changes, desertification and the Republic of Sudan', *GeoJournal* 33(4), 393–399

Ambler, J (1994) 'Small-scale surface irrigation in Asia: technologies, institutions and emerging issues', *Land Use Policy* 11(4), 262–274

Anderson, A B and Jardim, M A G (1989) 'Costs and Benefits of floodplain forest management by rural inhabitants in the Amazon estuary: a case study of Açai palm production', in J D Browder (ed) *Fragile Lands in Latin America: strategies for sustainable development*. Westview Press, Boulder, 114–129

Anderson, D M (1984) 'Depression, dust bowl, demography and drought: the colonial state and soil conservation in East Africa during the 1930s', *African Affairs* 83(232), 321–343

Anon (1982) 'Will polymers turn the desert green?', *New Scientist* 95(1316), p 305

Anon (1990) *Vetiver grass: a method of vegetative soil and water conservation* (3rd edn) World Bank, New Delhi

Anon (1996) 'Born of the Dust Bowl', *Government Executive* 28(6), 37

Anon (1996a) *The 1995–1996 Soil and Water Conservation Society Annual Report 'Toward the Next 50 Years'*. Special issue of the *Journal of Soil & Water Conservation* 51(5)

Anon (1996b) 'Urban agriculture a bountiful supplier', *BioCycle* 37(7), 8

Armillas, P (1971) 'Gardens on swamps', *Science* 174(4010), 653–661

Arnon, I (1972) *Crop Production in Dry Regions. Vol 1: background and principles.* Leonard Hill, London

Arnon, I (1981) *Modernization of Agriculture in Developing Countries: resources, potentials and problems.* Wiley, Chichester

Astatke, A Jutzi, S and Tedla, A (1989) 'Sequential cropping of vertisols in the Ethiopian highlands using a broadbed-and-furrow system', *ILCA Bulletin* 4, 15–20

Atampugre, N (1993) *Behind the Lines of Stone: the social impact of a soil and water conservation project in the Sahel.* Oxfam, Oxford

Balek, J (1977) *Hydrology and Water Resources in Tropical Africa.* Elsevier, Amsterdam

Barbier, E B and Bishop, J T (1995) 'Economic values and incentives affecting soil and water conservation in developing countries', *Journal of Soil & Water Conservation* 50(2), 133–137

Barker, G W W and Jones, G D B (1982) 'The UNESCO Libyan Valleys Survey VI: Investigation of a Romano-Libyan Farm, Part I', *Libyan Studies* 15(1), 1–70

Barker, T G jnr (1996) 'Environmental issues: an overview of tax policy', *Tax Adviser* 27(10), 602–604

Barrow, C J (1987) *Water Resources and Agricultural Development in the Tropics.* Longman, Harlow

Barrow, C J (1988) 'The present position and future development of rain-fed agriculture in the tropics', *Outlook on Agriculture* 17(3), 112–119

Barrow, E G G (1983) 'Use of micro-catchments for tree planting and soil conservation in semi-arid areas', in D B Thomas and W M Senga (eds) *Soil and Water Conservation in Kenya* (Proceedings of the Second National Workshop, Nairobi, 10–13 March 1982). Institute for Development Studies and Faculty of Agriculture, University of Nairobi (PO Box 30/97 Nairobi, Kenya), Occasional Paper no 42, 324–332

Bartolina, D (1996) 'Toward the next 50 years: excerpts from the 1995–1996 Soil and
Water Conservation Society Annual Report', *Journal of Soil & Water Conservation*
51(5), 448

Bateman, G H (1974) *A Bibliography of Low-Cost Water Technologies* (3rd edn).
Intermediate Technology Publications, London

Batterbury, S (1994) 'Soil and water conservation in Burkina Faso: the role of community organizations', *Appropriate Technology* 2(3), 6–9

Beach, T and Dunning, N P (1995) 'Ancient Maya terracing and modern conservation
in Peten rain forest of Guatamala', *Journal of Soil & Water Conservation* 50(2),
138–145

Beaumont, P (1971) 'Quanat systems in Iran', *Bulletin of the International
Association of Scientific Hydrology* 16(1), 39–50

Beaumont, P, Bonine, M E, McLachlan, A and McLachlan, K S (1989) *Quanat, Kariz
and Khettara: traditional water system in the Middle East and North Africa*.
University of London, Centre for Middle Eastern Studies Press, Middle East Centre,
SOAS, University of London, London

Bell, M and Roberts, N (1991) 'The political ecology of dambo soil and water-
resources in Zimbabwe', *Transactions of the Institute of British Geographers* 16(3),
301–318

Belsky, J M (1994) 'Soil conservation and poverty: lessons from upland Indonesia',
Society & Natural Resources 7(5), 429–443

Ben-Asher, J, Oron,G and Button, B J (1985) *Estimation of Runoff Volume for
Agriculture in Arid Lands*. Jacob Blaustein Institute for Desert Research, Ben
Gurion University of the Negev, Sede Boquer Campus

Benasalem, B (1981) 'Examples of soil and water conservation practices in North
African countries: Algeria, Morocco and Tunisia', in *FAO Soil Conservation and
Management in Developing Countries* (FAO Soils Bulletin No 33). FAO, Rome,
151–164

Benhur, M (1991) 'The effects of dispersants, stabilizer and slope length on runoff and
water harvesting farming', *Australian Journal of Soil Research* 29(4), 553–563

Benites, J R and Ofori, C S (1993) 'Crop production through conservation tillage –
effective tillage in the tropics', *Soil & Tillage Research* 27(1–4), 9–33

Bensalem, B (1981) 'Examples of soil and water conservation practices in North
African countries – Algeria, Morocco and Tunisia', in FAO (ed) *Soil Conservation
and Management in Developing Countries* (FAO Soils Bulletin No 33. FAO, Rome,
281–284

Bentley, J W (1987) 'Water harvesting on the Papago Reservation: experimental
agricultural technology in the guise of development', *Human Organization* 46(2),
141–146

Besteman, C (1996) '*Dhasheeg* agriculture in the Jubba Valley, Somalia', in J B Mabry
(ed) *Canals and Communities: small scale irrigation systems*. University of
Arizona Press, Tucson, 53–68

Bhusaan, L S, Yadev, R C, Singh, B L, Tiwari, A K, Singh, M, Gauer, M L and Ram, B
(1992) 'Prospects for rain-fed agriculture in gullied and ravine catchments through
soil and water conservation practices', *Journal of Arid Environments* 23(4),
433–441

Biamah, E K, Gichuki, F N and Kaumbutho, P G (1993) 'Tillage methods and soil
water conservation in eastern Africa', *Soil & Tillage Research* 27(1–4), 105–123

Biswas, A K (1990) 'Watershed management', *Water Resources Development* 6(4),
240–249

Biswas, A K and El-Habr, H N (1993) 'Environment and water resources management:
the need for a new holistic approach', *Water Resources Development* 9(2), 117–125

Blaikie, P (1985) *The Political Economy of Soil Erosion in Developing Countries*.
Longman, Harlow

Blaikie, P and Brookfield, H (1987) *Land Degradation and Society.* Methuen, London

Boardman, J, Foster, I D C and Dearing, J A (1990) *Soil Erosion on Agricultural Land.* Wiley, Chichester

Bochet, J J (1983) *Management of Upland Watersheds: participation of the mountain communities (FAO Conservation Guide – FAO Technical Paper No M–90).* FAO, Rome

Boers, Th M and Ben-Asher, J (1982) 'A review of rainwater harvesting', *Agricultural Water Management* 5(2), 145–158

Boers, Th M, Degraaf, M, Feddes, R A and Ben-Asher, J (1986a) 'A linear regression model combined with a soil water balance model to design micro-catchments for water harvesting in arid regions', *Agricultural Water Management* 11(3–4), 187–206

Boers, Th M, Zondervan, K and Ben-Asher, J (1986b) 'Micro-catchment water harvesting (MCWH) for arid zone development', *Agricultural Water Management* 12(1–2), 21–39

Bohra, D N and Issac, V C (1987) 'Runoff behaviour of soil sealants for harvesting rainwater in an arid environment', *Annals of Arid Zone* 26(3), 163–170

Bonine, M E (1996) 'Quanats and rural societies: sustainable agriculture and irrigation cultures in contemporary Iran', in J B Mabry (ed) *Canals and Communities: small scale irrigation systems.* University of Arizona Press, Tucson, 183–209

Bonnifield, P (1979) *The Dust Bowl: men, dirt and depression.* University of New Mexico Press, Albuquerque

Bonvallot, J (1986) 'Tabias et jessour du sud Tunisien: agricultur dans les zones marginales et parade a l'erosion', *Cahiers ORSTOM, se Pedologie* XXII(2), 163–171

Boserüp, E (1965) *The Conditions of Agricultural Growth.* Allen and Unwin, London

Boserüp, E (1965) *The Conditions of Agricultural Growth: the economics of agrarian change under population pressure.* Allen and Unwin, London (reprinted by Earthscan, London, 1993)

Bottrall, A (1993) 'Institutional aspects of watershed management', in ODA (ed) *Proceedings of a Conference on Priorities for Water Resources Allocation*, July 1992, Southampton Overseas Development Administration, London, 81–90

Brklacich, M, Bryant, C R and Smit, B (1991) 'Review and appraisal of the concept of sustainable food production systems', *Environmental Management* 15(1), 1–14

Bruins, H J, Evenari, M and Nessler, U (1986) 'Rainwater -harvesting agriculture for food production in arid zones: the challenges of African famine', *Applied Geography* 6(1), 13–32

Brunet, M (1990) 'Terrasses de cultures antiques: l'example de Delos, Cyclades', *Mediterranée* 71(1), 5–11

Brunner, U and Haefner, H (1986) 'The successful floodwater farming systems of the Sabeens, Yemen Arab Republic', *Applied Geography* 6(1), 77–86

Bryan, K (1926) 'Floodwater farming', *The Geographical Journal* 19(3), 444–456

Bunner, U (1986) 'The successful floodwater farming system of the Sabeans, Yemen Arab Republic', *Applied Geography* 6(1), 77–86

Busscher, W J, Reeves, D W, Kochhann, R A, Bauer, P J, Mullins, G L, Clapham, W M, Kemper, W D and Galerani, P R (1996) 'Conservation farming in southern Brazil: using cover crops to decrease erosion and increase infiltration', *Journal of Soil & Water Conservation* 51(3), 188–192

Butz, D (1989) 'The agricultural use of meltwater in Harpar Settlement, Pakistan', *Annals of Glaciology* 13(1), 35–39

Caponera, D (1973) *Water Laws in Moslem Countries (FAO Irrigation and Drainage Paper No 20/1).* FAO, Rome

Carrasco, D A and Witter, S G (1993) 'Constraints to sustainable soil and water conservation: a Dominican Republic example', *Ambio* XXII(6), 347–350

Carter, D C and Miller, S (1991) 'Three years' experience with an on-farm macro-catchment water harvesting system in Botswana', *Agricultural Water Management* 19(3), 191–203

Carter, M R (ed)(1994) *Conservation Tillage in Temperate Agroecosystems: development and application to soil, climate, and biological constraints*. CRC Press, Boca Raton

Chaker, M, El Abassi, H and Laouina, A (1996) 'Mountains, foothills and plains: investing in SWC in Morocco', in C Reij, I Scoones and C Toulmin (eds) *Sustaining the Soil: indigenous soil and water conservation in Africa*. Earthscan, London, 48–55

Chaudhry, A B (1982) 'Drawdown agriculture as a suitable substitute of traditional fadama land cultivation in the Kainji Lake Basin', in ILRI (ed) *Polders of the World* (vol II). International Institute for Land Reclamation and Improvement, Wageningen

Chleq, J L and Dupriez, H (1988) *Vanishing Land and Water, Soil and Water Conservation in Dry Lands* (Centre for Tropical Agriculture Handbook). Macmillan, London

Cohen, I, S, Lopes, V L, Slack, D C and Yanez, C H (1995) 'Assessing risk for water harvesting systems in arid environments', *Journal of Soil and Water Conservation* 50(5), 446–449

Collet, J and Boyd, J (1977) *Eight Simple Survey Levels* (Intermediate Technology Agricultural Equipment Leaflet No 42). Intermediate Technology Publications, London

Collins, J L (1986) 'Smallholder settlement of tropical South America: the social causes of ecological destruction', *Human Organization* 45(1), 1–10

Conelly, W T (1994) 'Population pressure, labor availability and agricultural disintensi-fication', *Human Ecology* 22(2), 145–170

Constantinesco, I (1981) Soil Conservation for Developing Countries. FAO, Rome

Cosgrove, D and Petts, G (eds)(1990) *Water, Engineering and Landscape*. Belhaven, London

Cowan, M (1982) *Flooded Water Meadows* (SWIAS Historical Monograph No 9). South Wiltshire Industrial Archaeology Society, Salisbury (UK)

Coward, E W (1979) 'Principles of social organisation in an indigenous irrigation system', *Human Organisation* 38(1), 28–36

Cracknell, J (1983) 'Soil and water conservation in the Windward Islands', *World Crops* 33(6), 218–220

Cressey, G B (1958) 'Quanats, karez and foggaras', *Geographical Review* 48(1), 27–44

Crews, T E and Gliessman, S R (1991) 'Raised field agriculture in Tlaxcala, Mexico: an ecosystem perspective on maintenance of soil fertility', *American Journal of Alternative Agriculture* 6(1), 9–17

Critchley, W R S, Reij, C, and Willcocks, T J (1994) 'Indigenous soil and water conser-vation: a review of the state of knowledge and prospects for building on traditions', *Land Degradation & Rehabilitation* 5(4),293–314

Critchley, W R S (1989) 'Building on a tradition of rainwater harvesting', *Appropriate Technology* 16(2), 10–12

Critchley, W R S (1990) 'How Africa's pastoralists catch the rain', *Ceres* 22(1), 41–45

Critchley, W R S (1991) *Looking After Our Land: new approaches to soil and water conservation in dryland Africa*. Oxfam, Oxford

Critchley, W R S (1991) *Looking After Our Land: soil and water conservation in dryland Africa*. International Institute for Environment and Development, London

Critchley, W R S, Reij, C and Seznec, A (1992) *Water Harvesting for Plant Production. Vol II: Case Studies and Conclusions for Sub-Saharan Africa* (World Bank Technical Paper, Africa Technical Department Series No 157). World Bank, Washington

Critchley, W R S, Reij, C P and Turner, S D (1992) *Soil and Water Conservation in Sub-Saharan Africa: towards sustainable production by the rural poor* (Report prepared for the International Fund for International Development (IFAD) by the Centre for Development Cooperation Services). Free University of Amsterdam, Amsterdam

Critchley, W R S, Reij, C and Wilcocks, T J (1994) 'Indigenous soil and water conservation: a review of the state of knowledge and prospects for building on traditions', *Land Degradation & Rehabilitation* 5(1), 293–314

Cross, M (1983) 'Last chance to save Africa's topsoil', *New Scientist* 199(1368), 288–293

Crosson, P (1995) 'Future supplies of water for world agriculture', in N Islam (ed) *Population and Food in the Early 21st Century: meeting future food demands of an increasing population*. International Food Policy Research Institute, Washington, 143–159

Cullis, A and Pacey, A (1992) *A Development Dialogue: rainwater harvesting in Turkana*. Intermediate Technology Publications, London

Dano, A M and Siapno, F E (1992) 'The effectiveness of soil conservation structures in steep cultivated mountain regions of the Philippines', in D E Walling, T R Davies and B Hasholt (eds) *Erosion, Debris Flows and Environment in Mountain Regions* (Proceedings of an International Symposium, Chengdu, 1992 – IAHS-AISH Publication No 209). International Association of Hydrological Sciences (IAHS), Wallingford, 399–405

Darch, J P (1988) 'Drained field agriculture in tropical Latin America: parallels from past to present', *Journal of Biogeography* 15(1), 87–95

De Fraiture, L (1991) 'Watering cans or channels? How assumptions on women's participation were swept away in the Senegal Valley', *At-Source* 19(4), 2–5

De Walt, B R (1994) 'Using indigenous knowledge to improve agriculture and natural resource management', *Human Organization* 53(2), 123–131

Debano, L F and Schmidt, L J (1990) 'Potential for enhancing riparian habitats in the southwestern United States with watershed practices', *Forest Ecology and Management* 33–34(1–4), 385–403

Dehn, M (1995) 'An evaluation of soil conservation techniques in the Ecuadorian Andes', *Mountain Research & Development* 15(2), 175–182

Denevan, W (ed) (1986) *The Cultural Ecology, Archaeology and History of terracing and Terrace Abandonment in the Colca Valley of Southern Peru*. University of Wisconsin Press, Madison

Dennell, R W (1982) 'Dryland agriculture and soil conservation: an archaeological study of check-dam farming and wadi siltation', in B Spooner and H S Mann (eds) *Desertification and Development: dryland ecology in social perspective*. Academic Press, London, 171–200

Denny, P (1993) 'Wetlands of Africa: introduction', in D F Whigham et al (eds) *Wetlands of the World vol 1: inventory, ecology and management*. Kluwer Academic, Dordrecht, 1–31

Dhar, S K (1994) 'Rehabilitation of degraded tropical forest watersheds with people's participation', *Ambio* XXIII(3), 216–221

Dharmasena, P B (1994) 'Conservation farming practices for small reservoir watersheds: a case study from Sri Lanka', *Agroforestry Systems* 28(3), 203–212

Dixey, F (1950) *A Practical Handbook of Water Supply* (2nd edn). Thomas Mabry, London

Donkin, R A (1978) *Agricultural Terracing in the Aboriginal New World* (Viking Fund Publication in Anthropology No 56). University of Arizona Press, Tucson

Doolette, J B and Smyle, J W (1990) *Soil and Moisture Conservation Technologies: a review of the literature* (World Bank Technical Paper No 127). World Bank, Washington

Doolittle, W E (1989) 'Arroyos and the development of agriculture in northern Mexico', in J O Browder (ed) *Fragile Lands of Latin America: strategies for sustainable development*. Westview Press, Boulder, 251–269

Douglas, T D, Kirkby, S J, Critchley, W R S and Park, G J (1994) 'Agricultural terrace abandonment in the Alpujarra, Andalucia, Spain', *Land Degradation & Rehabilitation* 5(4), 281–292

Dregne, H E (1986) 'Soil and water conservation – a global perspective', *Interçiencia* 11(4), 166–172

Drijver, C A and Marchand, M (1985) *Taming the Floods: environmental aspects of floodplain development in Africa.* Centre for Environmental Studies (CML), State University of Leiden, Leiden

Dupriez, H and De Leener, P (1992) *Ways of Water: run-off, irrigation and drainage* (Centre for Tropical Agriculture Handbook) (translated from Dutch by B O'Meara). Macmillan, London

Duvdevani, S (1957) 'Dew research for arid agriculture', *Discovery* 18, 330–334

Easter, K A, Dixon, J A and Huffschmidt, M M (eds) (1986) *Watershed Resource Management: an integrated framework with studies from Asia and the Pacific.* Westview Press, Boulder

Eckholm, E P (1976) *Losing Ground: environmental stress and world food production.* Pergamon, Oxford

Eder, J F (1982) 'No water in the terraces: agricultural stagnation and social change at Banaue, Ifugao', *Philippine Quarterly of Culture and Society* 10(3), 101–106

Edwards, D T (1995) 'Small farmers' protection of the watersheds: the experience since the 1950s', in M J Gargett (ed) *Development Projects: issues for the 1990s* (vol 1), Proceedings of the 25th Annual Conference, 6–8 April 1995. Development Project Planning Centre, University of Bradford, Bradford (UK)

El Amami, S (1977) 'Traditional technologies and development of the African environments: utilisation of runoff waters; the meskats and other techniques of Tunisia', *African Environment* 3(3–4), 107–120

El Kassas, M (1979) 'Barren answers', *Development Forum* VII(3) 6

El Wakeel, A and Astatke, A (1995) 'Intensification of agriculture on Vertisols to minimize land degradation in parts of the Ethiopian highlands', *Land Degradation & Rehabilitation* 7(1), 57–67

Elliott, L F and Wildung, R E (1992) 'What biotechnology means for soil and water conservation', *Journal of Soil & Water Conservation* 47(1), 17–20

Ellis-Jones, J and Simms, B (1995) 'An appraisal of soil conservation techniques in Honduras, Mexico and Nicaragua', *Project Appraisal* 10(2), 125–134

Emmerich, W E Frasier, G W and Fink, D H (1987) 'Relations between soil properties and effectiveness of low-cost water harvesting treatments', *Soil Science Society of America Journal* 51(1), 213–219

Ericksson, C L and Candler, K L (1989) 'Raised fields and sustainable agriculture in the Lake Titicaca Basin of Peru', in J D Browder (ed) *Fragile Lands of Latin America: strategies for sustainable development.* Westview Press, Boulder, 230–249

Erskine, J M (1992) 'Vetiver grass: its potential use in soil and moisture conservation in Southern Africa', *South African Journal of Science* 88(6), 298–299

Ervine, D E (1994) 'Soil and water conservation down on the farm: a changing economic landscape', *Journal of Soil & Water Conservation* 49(3), 232–234

Evenari, M, Nessler, U, Rogel, A and Shenk, O (eds) (1982) *Fields and Pastures in Deserts: a low cost method for agriculture in semi-arid lands.* Eduard Roether, Buchdruckerei und Verlag, Darmstadt

Evenari, M, Shanan, L and Tadmore, N H (1971) *The Negev: the challenge of a desert.* Harvard University Press, Cambridge

Ewell, P, Hughes, D, Sanders, D W, Gallagher, R, Ransom, J K, Wood, A and Wortmann, C S (1993) 'Current aid agency approaches to soil and water conservation', in N W Hudson and R Cheatle (with A Wood and F Gichuki) (eds) *Working with Farmers for Better Land Husbandry.* Intermediate Technology Publications, London, 45–53

FAO (1977) *Conservation Guide 1: guidelines for watershed management*. FAO, Rome

FAO (1977) *Guidelines for Watershed Management* (FAO Conservation Guide No 1). FAO, Rome

FAO (1987) *Soil and Water Conservation in Semi-Arid Areas* (FAO Soils Bulletin No 57). FAO, Rome

FAO (1987) *Spate Irrigation Proceedings of the Subregional Expert Consultation on Wadi Development for Agriculture in the Northern Yemen, Aden* (PDR Yemen, 6–10 December, 1987) (Working Paper No AG:UNDP/RAB/840/30). FAO, Rome

FAO (1988) *Soil and Water Conservation in Semi-Arid Areas*. FAO, Rome

FAO (1988) *Watershed Management Field Manual: slope treatment measures and practices* (FAO Conservation Guide 13/3) FAO, Rome

FAO (1991) *A Study of the Reasons for Success or Failure of Soil Conservation Projects* (FAO Soils Bulletin No 64) FAO, Rome

FAO (1996) *Environmental Impact Assessment of Irrigation and Drainage Projects* (FAO Irrigation and Drainage Paper No 53). FAO, Rome

Farrington, I S (1980) 'The archaeology of irrigation canals, with special reference to Peru', *World Archaeology* 11(3), 287–305

Farrington, J and Lobo, C (1997) *Scaling up Participatory Watershed Development in India: lessons from the Indo-German Watershed Development Programme* (ODI Natural Resource Perspectives No 17). Overseas Development Institute, London

Faulkner, H (1995) 'Gully erosion associated with expansion of unterraced almond cultivation in the coastal Sierra de Lujar, southern Spain', *Land Degradation & Rehabilitation* 6(3), 179–200

Fink, D H (1984) 'Paraffin-wax water harvesting soil treatment improved with antistripping agents', *Soil Science* 138(1), 46–53

Finkel, H J (1986) *Semiarid Soil and Water Conservation*. CRC Press, Boca Raton

Foster, N R (1992) 'Protecting fragile lands: new reasons to tackle old problems', *World Development* 20(4), 571–585

Francis, D G (1994) *Family Agriculture: tradition and transformation*. Earthscan, London

Frasier, G W (1975) 'Water harvesting for livestock, wildlife, and domestic use', in USDA (ed) *Proceedings of the Water Harvesting Symposium: ARS W–22, 1975*. US Water Conservation Laboratory (Science and Education Division), US Department of Agriculture, Phoenix, 40–49

Frasier, G W (1984) *Water Harvesting: including new techniques of maximising rainfall in semi-arid areas* (Proceedings of the Fourth Agricultural Sector Symposium). World Bank, Washington

Frasier, G W and Myers, L E (1983) *Handbook of Water Harvesting* (USARC Agricultural Handbook No 600). US Department of Agriculture, Agricultural Research Service, Washington

Frasier, G W, Dutt, G R and Fink, D H (1987) 'Sodium- salt treated catchments for water harvesting', *Transactions of the American Society of Agricultural Engineers* (ASCE) 30(3), 658–664

Freebairn, D M, Loch, R J and Cogle, A L (1993) 'Tillage methods and soil conservation in Australia', *Soil & Tillage Research* 27(1–4), 303–325

Furguson, B K (1996) 'Estimation of direct runoff in the Thornthwaite water balance', *Professional Geographer* 48(3), 263–271

Furon, R (1963) *The Problem of Water: a world study*. Faber and Faber, London

Gallart, F, Llorens, P and Latron, J (1994) 'Studying the role of old agricultural terraces on runoff generation in a small Mediterranean mountainous basin', *Journal of Hydrology* 159(1–4), 291–303

Garcia-Perez, J D, Charlton, C and Ruiz, P M (1995) 'Landscape changes as visible indicators in the social, economic and political process of soil erosion: a case study of the municipality of Puebla de Valles (Guadalajara Province), Spain', *Land Degradation & Rehabilitation* 6(3), 149–162

Gerasimenko, V P (1992) 'Predictions of soil and water conservation efficiency of methods for cultivating slope land', *EurAsian Soil Science* 24(5), 69–82

Ghimire, K B (1993) *Linkages between Population, Environment and Development.* United Nations Research Institute for Social Development, Geneva

Gilbertson, D D (ed) (1986) 'Runoff (floodwater) farming and rural water supply in arid lands', *Applied Geography* 6(1), 5–11

Gilbertson, D D and Chisholm, N W T (1996) 'UVLS XXVIII – manipulating the desert environment: ancient walls, floodwater farming and territoriality in the Tripolitanian pre-desert of Libya', *Libyan Studies* 27(1), 17–52

Gilbertson, D D and Hunt, C O (1990) 'ULVS XX – geomorphological studies of the Romano-Libyan farm, its floodwater control structures and weathered building stone at site LM4, at the confluence of Wadi el Amud and Wadi el Bagul in the Libyan pre-desert', *Libyan Studies* 21(1), 25–42

Gindel, I (1965) 'Irrigation of plants with atmospheric water within the desert', *Nature* 207(5002), 1173–1175

Giraldez, J V, Ayuso, J L, Garcia, A Lopez, J G and Roldan, J (1988) 'Water harvesting strategies in the semiarid climate of southeast Spain', *Agricultural Water Management* 14(1–4), 253–263

Gischeru, P T (1994) 'Effects of residue mulch and tillage on soil moisture and conservation', *Soil Technology* 7(3), 209–220

Gischler, C E (1979) *Water Resources in the Arab Middle East and North Africa.* Middle East and North African Studies Press, Cambridge

Gischler, C E (1982) 'Helpful clouds', *Development Forum* 10(2), 7

Gischler, C E and Jáuregui, C F (1984) 'Low-cost techniques for water conservation and management in Latin America', *Nature and Resources* XX(3), 11–18

Glantz, M H (1992) 'Global warming and environmental change in sub-Saharan Africa', *Global Environmental Change* 2(3), 183–204

Glantz, M H (ed) (1994) *Drought Follows the Plough: cultivating marginal areas.* Cambridge University Press, Cambridge

Goudie, A S (ed) (1990) *Techniques for Desert Reclamation.* Wiley, Chichester

Goudie, A S and Wilkinson, J (1977) *The Warm Desert Environment.* Cambridge University Press, Cambridge

Gould, J E (1994) 'Rainwater catchment systems technology: recent developments in Africa and Asia', *Science, Technology and Development* 12(2–3), 24–39

Govindasamy, R (1997) 'Simulation of an irrigation tank for modernization', *Water Resources Development* 7(2), 97–106

Govidrasamy, R and Balasubramaniam, R (1990) 'Tank irrigation in India: problems and prospects', *International Journal of Water Resources Development* 6(3), 211–217

Greenland, D J and Lal, R (1977) *Soil Conservation and Management in the Humid Tropics.* Wiley, Chichester

Greenwood, M N (1986) 'Extension's role in promoting soil and water conservation', *Journal of Soil & Water Conservation* 41(1), 4

Grepperud, S (1995) 'Soil conservation and governmental policies in tropical areas: does aid worsen the incentives for arresting erosion? ', *Agricultural Economics* 12(2), 129–140

Grimshaw, R G and Helfer, L (eds)(1995) *Vetiver Grass for Soil and Water Conservation, Land Rehabilitation, and Embankment Stabilization: a collection of papers and newsletters compiled by the Vetiver Network* (World Bank Technical Paper No 273). World Bank, Washington

Groenfeldt, D (1991) 'Building on tradition: indigenous irrigation knowledge and sustainable development in Asia', *Agriculture and Human Values* 8(1), 114–120

Grove, A T (1993) 'Water use by the Chagga on Kilimanjaro', *African Affairs* 92(368), 421–448

Grove, A T (ed) *The Niger and its Neighbours: environmental history and hydrology, human use and health hazards of the major west African rivers.* A A Balkema, Rotterdam

Grove, R H (1995) *Green Imperialism: colonial expansion, tropical island edens and the origins of environmentalism, 1600–1860.* Cambridge University Press, Cambridge

Guillet, D (1987) 'Terracing and irrigation in the Peruvian Highlands', *Current Anthropology* 28(4), 409–430

Gumbs, F A (1993) 'Tillage methods and soil and water conservation methods in the Caribbean', *Soil & Tillage Research* 27(1–4), 341–354

Gupta, G N (1989) 'Integrated effect of water harvesting, manuring and mulching on soil properties, growth and yield of crops in pearl millet – mungbean rotation', *Tropical Agriculture* 66(3), 233–239

Gupta, G N (1994) 'Influence of rain water harvesting and conservation practices on growth and biomass production of *Azadirachta indica* in the Indian Desert', *Forest Ecology and Management* 70(1–3), 329–339

Gupta, G N (1995) 'Rain-water management for tree planting in the Indian Desert', *Journal of Arid Environments* 31(2), 219–235

Haagsma, B (1995) 'Traditional water management and state intervention: the case of Santo Antão, Cape Verde', *Mountain Research & Development* 15(1), 39–56

Haagsma, B and Reij, C (1993) '*Frentes-de-trabalho*: potentials and limitations of large-scale labour employment for soil and water conservation in Cape Verde', *Land Degradation & Rehabilitation* 4(2), 73–85

Hagen, I J (1991) 'A wind erosion prediction system to meet user needs', *Journal of Soil & Water Conservation* 46(1), 106–111

Hagmann, J (1996) 'Mechanical soil conservation with contour ridges: cure for, or cause of, rill erosion?', *Land Degradation & Development* 7(2), 143–160

Hagmann, J, and Muwira, K (1996) *Indigenous soil and water conservation in southern Zimbabwe: a study on technologies, historical changes and recent developments under participatory research and extension* (IIED Issue Paper No 63). International Institute for Environment and Development, London

Halbach, D W, Runge, C F and Larson, W E (1988) *Making Soil and Water Conservation Work: scientific and policy perspectives.* Soil Conservation Society of America, Ankeny

Hall, A E, Cannel, G H and Lawton, H W (eds)(1979) *Agriculture in Semi-Arid Environments* (Ecological Studies: analysis and synthesis, vol 34). Springer-Verlag, Berlin

Hall, A L (1979) *Drought and Irrigation in North East Brazil.* Cambridge University Press, Cambridge

Hallsworth, E G (1997) *Anatomy, Physiology and Psychology of Erosion.* Wiley, New York

Hanrahan, M S and McDowell, W (1997) 'Policy variables and program choices: soil and water conservation results from the Cochabamba high valleys', *Journal of Soil & Water Conservation* 52(4), 252–259

Harden, C P (1992) 'A new look at soil erosion processes on hillslopes in highland Ecuador', in D E Walling, T R Davies and B Hasholt (eds) *Erosion, Debris Flows and Environment in Mountain Regions* (Proceedings of an International Symposium, Chengdu, 1992 – IAHS-AISH Publication No 209). International Association of Hydrological Sciences (IAHS), Wallingfotd, 77–85

Hastorf, C (1989) 'Agricultural production in the Central Andes: lost strategies', in C Gradwin and K Truman (eds) *Food and Farm: current debates and policies* (Monographs in Economic Anthropology No 7). University Press of America, New

York, 231–253

Haverkort, B, Van der Kemp, J and Waters-Bayer, A (eds) (1991) *Joining Farmers' Experiments: experiences in participatory technology development.* Intermediate Technology Publications, London

Helms, D and Flader, S L (1985) 'The history of soil and water conservation: a symposium: introduction', *Agricultural History* 59(2), 103–106

Hervé, D, Poupon, H and Rousseau, P (1989) 'Irrigation et maitrise de l'eau sur un versant des Andes Peruviennes', *Etudes Rurales* 115 and 116, 159–179

Hien, F G, Rietkerk, M and Stroosnijder, L (1997) 'Soil variability and the effectiveness of soil and water conservation in the Sahel', *Arid Soil Research and Rehabilitation* 11(1), 1–8

Hillel, D (1994) *Rivers of Eden: the struggle for water and the quest for peace in the Middle East.* Oxford University Press, Oxford

Hillman, F (1980) *Water Harvesting in Turkana District* (ODI Agricultural Administration Unit, Pastoral Network Paper No 10d). Overseas Development Institute, London

Hilversum, L (1992) *The Terraces of Wallo: a report on the Soil Conservation Programme in the Barkena Catchment, Ethiopia* (Nordic Conference on Environment and Development, Stockholm, 1997). Panos, London

Hinchcliffe, F, Guijt, I, Pretty, J N and Shah, P (1995) *New Horizons: the economic, social and environmental impacts of participatory watershed development* (IIED Gatekeeper Series No SA50). International Institute for Environment and Development, London

Hoag, D, Lilley, S, Cook, M and Wright, J (1988) 'Extension's role in soil and water conservation', *Journal of Soil & Water Conservation* 43(2), 126–129

Hogg, R (1988) 'Water harvesting and agricultural production in semi-arid Kenya', *Development & Change* 19(1), 69–87

Holloway, J E and Guy, D C (1990) 'Coordination of state soil and water conservation and farmland regulatory programs', *Journal of Soil & Water Conservation* 45(5), 543–546

Hotchkiss, P and Lambert, B (1987) *Dambos and micro-scale irrigation: technical and social aspects in Zimbabwe.* Overseas Development Institute, Irrigation Management Network, 11 December 1987. ODI, London

Hubbard, A J and Hubbard, G (1907) *Neolithic Dew Ponds and Cattleways* (2nd edn). Longmans Green & Co, London

Hudson, N and Cheatle, R (with Wood, A and Gichuki, F) (eds)(1993) *Working with Farmers for Better Land Husbandry.* Intermediate Technology Publications, London

Hudson, N W (1987) *Soil and Water Conservation in Semi-arid Areas* (FAO Soils Bulletin No 57). FAO, Rome

Hudson, N W (1992) *Land Husbandry* (9th edn). Batsford, London

Hudson, N W and Cheattle, R (with A Wood and F Gichuki) (eds) (1993) *Working with Farmers for Better Land Husbandry.* Intermediate Technology Publications, London

Hulme, M (1992) 'Rainfall changes in Africa: 1931–1960 to 1961–1990', *International Journal of Climatology* 12(7), 685–699

Hulme, M (1994) 'Using climatic information in Africa: some examples related to drought, rainfall forecasting and global warming', *IDS Bulletin* 25(2), 59–68

Hulugalle, N R, De Koning, J and Matlan, P J (1990) 'Effect of rock bunds and tied ridges on soil water content and soil properties in the Sudan savannah of Burkina Faso', *Tropical Agriculture* 67(2), 149–153

Huszar, P C and Cochrane, H C (1990) 'Constraints to conservation farming in Java's uplands', *Journal of Soil & Water Conservation* 45(3), 420–423

Hutchinson, C F, Dutt, G R and Garduno, A M (eds) (1981) *Rainfall Collection for Agriculture in Arid and Semiarid Regions* (Proceedings of a Workshop at the University of Arizona). Commonwealth Agricultural Bureaux, Slough

IDRC Review of Research for Development (1997–1998) 34–36

IFAD (1986) *Soil and Water Conservation in Sub-Saharan Africa: issues and options.* (Centre for Development Cooperation Services, Free University, Amsterdam) International Fund for Agricultural Development (IFAD), Rome

Intermediate Technology Development Group (1969) *The Introduction of Rainwater Catchment Tanks and Micro-irrigation to Botswana.* ITDG, London

IUCN (1982) *Convention de Ramsar/Ramsar Convention 1982 Conférence extraordinaire des parties contractantes, 2–3 December 1982, Paris, France.* IUCN, Gland

IUCN (1997) *Indigenous Peoples and Sustainability: cases and actions* (IUCN Indigenous Peoples and Conservation Initiative, Inter-Commission Task Force on Indigenous Peoples). International Books, Utrecht

Jackson, I L (1977) *Climate, Water and Agriculture in the Tropics.* Longman, London

Jeffords, J (1987) 'Soil and water conservation: from here to where? ', (editorial) *Journal of Soil & Water Conservation* 42(6), p 414

Jiménez-Osornio, J J and Gomez-Pompa, A (1991) 'Human role in shaping of the flora in a wetland community, the chinampa', *Landscape and Urban Planning* 20(1–3), 47–51

Jodha, N S (1990) 'Mountain agriculture: the search for sustainability', *Journal of Farming Systems Research–Extension* 1(1), 55–77

Johnson, D L and Lewis, L A (1995) *Land Degradation: creation and destruction.* Blackwell, Oxford

Jolly, R W, Elveld, B,McGramm, V and Raitt, D (1985) 'Transferring soil conservation to farmers', in R F Follett and B A Stewart (eds) *Soil Erosion and Crop Productivity.* American Society of Agronomy/Soil Science Society of America/Crop Science Society of America, Madison, 459–480

Jones, J A A (1997) *Global Hydrology: processes, resources and environmental management.* Addison Wesley Longman, Harlow

Jumkis, A R (1965) 'Aerial-wells: secondary sources of water', *Soil Science* 100(2), 83–95

Jurion, R and Henry, J (1969) *Can Primitive Farming be Modernized?* INEAC, Brussels

Juvic, J O, Dos Anjos, C and Nullet, D (1995) 'Direct cloud water recovery by inertial impaction: implications for large scale water supply in Cape Verde Islands', *Theoretical & Applied Climatology* 51(1–2), 89–96

Kahlown, M A and Hamilton, J R (1994) 'Status and prospects of karez irrigation', *Water Resources Bulletin* 30(1), 125–148

Kahlown, M A and Hamilton, J R (1996) 'Sailaba irrigation practices and prospects', *Arid Soil Research and Rehabilitation* 10(2), 179–191

Kaihura, F B S and Mowo, J G (1993) 'Soil and water conservation in Tanzania', in N Hudson and R Cheatle (with A Wood and F Gichuki) (eds) *Working With Farmers for Better Land Husbandry.* Intermediate Technology Development Group, London, 29–33

Kalitsi, E K (1973) 'Volta Lake in relation to human population and some issues in economics and management', in W C Ackermann, G F White and E B Worthington (eds) *Man-Made Lakes: their problems and environmental effects* (AGU Monograph No 17). American Geophysical Union, Washington, 77–85

Kamra, S K, Narayana, V V D and Rao, K V G K (1986) 'Water harvesting for reclaimed alkali soils', *Agricultural Water Management* 11(2), 127–135

Kang, B T, Wilson, G F and Lawson, T L (1984) *Alley Cropping: a stable alternative to shifting agriculture.* IITA, Ibadan (Nigeria)

Kasivelu, S, Howes, R and Devavaram, J (1995) *The Uses of PRA in Rehabilitating Minor Irrigation Tanks* (IIED PLA Notes No 22). International Institute for Environment and Development, London

Kaushik, S K and Gautam, R C (1994) 'Response of rain-fed pearl-millet (*Pennisetum glaucum*) to water harvesting, moisture conservation and plant population in light soils', *Indian Journal of Agricultural Sciences* 64(12), 858–860

Kavalo, S and Nehanda, G (1993) 'Soil and water conservation in Tanzania', in N Hudson and R Cheatle (with A Wood and F Gichuki) (eds) *Working with Farmers for Better Land Husbandry*. Intermediate Technology Publications, London, 41–44

Kerr, J and Sanghi, N K (1992) *Indigenous Soil and Water Conservation in India's Semi-Arid Tropics* (Gatekeeper Series No 34). International Institute for Environment and Development (IIED), London

Khan, M F K and Nawaz, M (1995) 'Karez irrigation in Pakistan', *GeoJournal* 37(1), 91–100

Khasiani, S A J (1992) 'Women in soil and water conservation projects – an assessment', in S A J Khasiani (ed) *Groundwork: African women as environmental managers*. African Centre for Technology Studies, Nairobi, 27–40

Kheating, M (1993) *The Earth Summit's Agenda for Change: a plain language version of Agenda 21*. Centre for Our Common Future, Palais Wilson, 52 Rue des Pâquis, Geneva CH–1201

Kidme, R M and Stocking, M (1995) 'Rationality of farmer perceptions of soil erosion: the effectiveness of soil conservation in semi arid Kenya', *Global Environmental Change* 5(4), 281–295

Kiepe, P (1995) *No Runoff, No Soil Loss: soil and water conservation in hedgerow barrier systems* (Tropical Resource Management Papers No 10) (summaries in English and Dutch). Office for International Relations, the Agricultural University, Wageningen

Kiome, R M and Stocking, M (1995) 'Rationality of farmer perception of soil erosion: the effectiveness of soil conservation in semiarid Kenya', *Global Environmental Change: Human and Policy Dimensions* 5(4), 281–295

Kirschenmann, F (1992) 'Green vs gene', *Agricultural Engineering* 73(1), 22–24

Kobori, I and Glantz, M H (eds)(1998) *Shrinking Seas: central Eurasian water crisis*. United Nations University Press, Tokyo

Kogan, F and Sullivan, J (1993) 'Development of global drought-watch system using NOAA/avhrr data', *Advances in Space Research* 13(5), 219–222

Kolakar, A S, Murthy, K N K and Singh, N (1983) '"Khadin": a method of harvesting water for agriculture in the Thar Desert', *Journal of Arid Environments* 6(1), 59–66

Kottman, R M (1984) 'A real-estate tax levy for soil and water conservation', *Journal of Soil & Water Conservation* 39(5), p 284

Krantz, B A (1979) *Small Watershed Development for Increased Food Production*. International Centre for Research in the Semi-Arid Tropics (ICRISAT), Hyderabad

Kronen, M (1994) 'Water harvesting and conservation techniques for smallholder crop production systems', *Soil & Tillage Research* 32(1), 71–86

Kutsch, H (1982) 'Principal Features of Water Concentrating Culture', *Trier Geographical Studies* no 5. Trier University, Trier (Germany)

Kutsch, H (1982) *Principal Features of a Form of Water Concentrating Culture* (Trier Geographical Studies no 5 – English translation). Universitat Trier, Trier (Germany)

Kutsch, H (1983) 'Currently used techniques in rain-fed water concentrating culture: the example of the Anti-Atlas', *Applied Geography and Development* 21(3), 114–125

Lafond, G P, Derksen, D A, Loepphy, H A and Struthers, D (1994) 'An agronomic evaluation of conservation-tillage systems and continuous cropping in east-central Saskatchewan', *Journal of Soil & Water Conservation* 49(4), 387–393

Lal, R (1983) *No-till Farming: soil and water conservation and management in the humid and subhumid tropics* (IITA Monograph No 2). International Institute for Tropical Agriculture, Ibadan

Lal, R (1990) 'Low resource agriculture alternatives in sub-Saharan Africa', *Journal of Soil & Water Conservation* 45(4), 437–445

Lal, R (1991a) 'Soil structure and sustainability', *Journal of Sustainable Agriculture* 1(1), 67–92

Lal, R (1991b) 'Tillage and agricultural sustainability', *Soil & Tillage Research* 20(2–4), 133–146

Lal, R (1994) 'Water management in various crop production systems related to soil tillage', *Soil & Tillage Research* 30(2–4), 169–185

Lal, R and Russel, E W (eds)(1981) *Tropical Agricultural Hydrology*. Wiley, Chichester

Lambert, R A, Hotchkiss, P F, Roberts, N, Faulkner, R D, Bell, M and Windram, A (1990) 'Use of wetlands (dambos) for micro-scale irrigation in Zimbabwe', *Irrigation and Drainage Systems* 4(1), 17–28

Lambert, R and Faulkner, R (1989) 'Simple irrigation technology for micro-scale irrigation', *Waterlines* 7(4), 26–28

Langworthy, M and Finant, T J (1997) *Waiting for Rain: agriculture and ecological imbalance in Cape Verde*. Lynne Rienner, London

Laryea, K B (1992) 'Rainfed agriculture: water harvesting and soil water conservation', *Outlook on Agriculture* 21(4), 271–277

Lavee, H, Poesen, J and Yair, A (1997) 'Evidence of high efficiency water harvesting by ancient farmers in the Negev Desert, Israel', *Journal of Arid Environments* 35(2), 341–348

Lawrence, P (1986) 'Spate irrigation in the Yemen Arab Republic', *ODU Bulletin* 7–10 April 1986

Layzell, D (1968) *Construction and Use of the Line Level*. Kenya Department of Agriculture, Nairobi

Le Houérou, H N and Lundholm, B (1976) 'Complimentary activities for the improvement of the economy and the environment in marginal drylands', in A Rapp, H N LeHouérou and B Lundholm (eds) *Can Desert Encroachment be Stopped? A Study with Emphasis on Africa*. Almqvist and Wiksell, Stockholm, 217–229

Le Moigne, G, Barghouti, S and Plusquellec, H (eds)(1989) *Technological and Institutional Innovation in Irrigation* (World Bank Technical Paper No 94). World Bank, Washington

Leach, M and Mearns, R (1996) *The Lie of the Land: challenging received wisdom on the African environment*. James Currey, London

Leblond, B and Guérin, L (1988) *Soil Conservation: project design and implementation using labour-intensive techniques* (2nd edn). International Labour Organization, Geneva

Lehmann, R (1993) 'Terrace degradation and soil erosion on Naxos Island, Greece', in S Wicherek, (ed) *Farm Land Erosion in Temperate Plains Environments and Hills* (Proceedings of an International Symposium on Farm Land Erosion, École Normale Superieure Fontenay-Saint Cloud, Paris, 25–29 May 1992). Elsevier Science, Amsterdam, 429–450

Lewis, J (1984) *Baringo Pilot Semi-Arid Area Project* (Summary of Interim Report). Republic of Kenya, BPSAAP, PO Marigat, via Nakuru, Kenya (mimeo)

Lewis, L A (1992) 'Terracing and accelerated soil loss on Rwandian steeplands: a preliminary investigation of the implications of human activities affecting soil moisture', *Land Degradation & Rehabilitation* 3(4), 241–246

Lewis, L A and Nyamulinda, V (1996) 'The critical role of human activities in land degradation in Rwanda', *Land Degradation & Rehabilitation* 6(1), 47–56

Lightfoot, D R (1993) 'The cultural-ecology of Puebloan pebble-mulch gardens', *Human Ecology* 21(2), 115–143

Lightfoot, D R (1994) 'Morphology and ecology of lithic-mulch agriculture', *Geographical Review* 84(2), 172–185

Lightfoot, D R (1996) 'Moroccan *khettara*: traditional irrigation and progressive desiccation', *Geoforum* 27(2), 261–273

Lima, R R (1956) *A Agricultura nas Várzeas do Estuario do Amazonas* (Boletim Técnico do Instituto Agronômico do Norte No 33). Instituto Agronômico do Norte, Belém

Lindstrom, M J (1986) 'Effects of residue harvesting on water runoff, soil erosion and nutrient loss', *Agricultural Economics & Environment* 16(2), 103–112

Little, P D and Horowitz, M M (eds) (1987) *Lands at Risk in the Third World: local-level perspectives.* Westview Press, Boulder

Lopes, V L and Meyer, J (1993) 'Watershed management program on Santiago Island, Cape Verde', *Environmental Management* 17(1), 51–57

Lopez, M V, Arrue, J L and Sanchezgiron, V (1996) 'A comparison between seasonal changes in soil water storage and percolation resistance under conventional and conservation tillage systems in Aragon', *Soil & Tillage Research* 37(4), 251–271

Lovenstein, H M, Berliner, P R and Van Keulen, H (1991) 'Runoff agroforestry in arid lands'. *Forest Ecology and Management* 45(1–4), 59–70

Lutz, E, Pagiola, S and Reiche, C (1994a) 'The costs and benefits of soil conservation: the farmers' viewpoint', *World Bank Research Observer* (US) 9(2), 273–295

Lutz, E, Pagiola, S and Reiche, C (eds) (1994b) *Economic and Institutional Analysis of Soil Conservation Projects in Central America and The Caribbean* (World Bank Environment Paper No 8). World Bank, Washington

Maaren, H and Dent, M (1995) 'Broadening participation in integrated catchment management for sustainable water resources development', *Water Science and Technology* 32(5–6), 161–167

Mabry, J B (ed) (1996) *Canals and Communities: small-scale irrigation systems.* University of Arizona Press, Tucson

Magrath, W B (1990) *Economic Analysis of Soil Conservation Technologies* (World Bank Technical Paper No 127). World Bank, Washington

Mahendrarajah, S, Jakeman, A J and Young, P C (1996) 'Water supply in monsoon Asia: modelling and predicting small tank storage', *Ecological Modelling* 84(1–3), 127–137

Mallappa, M, Radder, G D and Manyunath, S (1992) 'Growth and yield of *rabi* sorghum as influenced by soil and water conservation', *Annals of Arid Zone* 31(1), 37–43

Mandal, B K, Saha, A, Dhara, M C and Bhunia, S R (1994) 'Effect of zero tillage and conventional tillage on winter oil seed crops in West Bengal', *Soil & Tillage Research* 29(1), 49–57

Mannasah, J T and Briskey, E J (eds)(1981) *Advances in Food Producing Systems in Arid and Semi-Arid Lands* (Part A). Academic Press, London

Mannering, J V and Fenster, C R (1983) 'What is conservation tillage? ', *Journal of Soil & Water Conservation* 38(3), 141–143

Manrique, L A (1993) 'Soil management and conservation in the tropics: indigenous and adopted technology', *Communications in Soil Science & Plant Analysis* 24(13–14), 1617–1644

Margaris, N S (1987) 'Desertification in the Aegean Islands', *Ekistics* No 323–324, 132–136

Margaris, N S (1992) 'Primary sector and environment in the Aegean Islands, Greece', *Environmental Management* 16(5), 569–574

Margraf, J, Voggelsberger, M and Milan, P P (1996) 'Limnology of Ifugao rice terraces, Philippines', in F Schiemer and K T Boland (eds) *Perspectives in Tropical Limnology.* SPB Academic, Berlin, 305–319

Martin, E D and Yoder, R (1987) *Institutions for Irrigation Management in Farmer-Managed Systems: examples from the hills of Nepal* (IIMI Research Paper No 5). International Irrigation Management Institute, Sri Lanka

Martinez, T L (1990) 'L'agriculture en terrasses dans les Pyrenées centrales espagnoles', *Mediterranée* 71(1), 37–42

Matheny, R T (1982) 'Ancient lowland and highland Maya water and soil conservation strategies', in K V Flannery (ed) *The Early Mesoamerican Village*. Academic Press, London, 157–203

Matlock, W G and Dutt, G R (1984) *A Primer on Water Harvesting and Runoff Farming*. Irrigation and Water Management Institute, College of Agriculture, University of Arizona, Tucson

Mellerowicz, K T, Rees, H W, Chow, T L and Ghanem, I(1994) 'Soil conservation planning at the watershed level using the Universal Soil Loss Equation with GIS and microcomputer technologies: a case study', *Journal of Soil & Water Conservation* 49(2), 194–200

Michaelson, T (1991) 'Participatory approaches in watershed management', *Unasylva* 164(42), 3–7

Mitsch, W J and Grosselink, J G (1993) *Wetlands* (2nd edn). Van Nostrand Reinhold, New York

Mock, J F (1985) *Traditional Irrigation Schemes and Potential for their Improvement* (Irrigation Symposium, Water Congress, Berlin, 1985). P Parey, Hamburg

Mohamed, Y A (1993) 'Water harvesting in Darfur, Sudan', in N Hudson, R W Cheatle, A Wood and F Gichuki (eds) *Working with Farmers for Better Land Husbandry*. Intermediate Technology Publications, London, 169–171

Moldenhauer, W C and Hudson, N W (eds) (1988) *Conservation Farming on Steep Lands*. Soil and Water Conservation Society, Ankeny

Moldenhauer, W C, Hudson, N W and Sheng, T C (eds)(1991) *Development of Conservation Farming on Hillslopes*. Soil and Water Conservation Society, Ankeny

Mondal, R C (1974) 'Pitcher farming', *Appropriate Technology* 1(3), 7

Monteith, J L (1963) 'Dew facts and fallacies', in A J Rutter and F H Whitehead (eds) *The Water Relations of Plants*. Blackwell Scientific, London, 37–56

Moreno, F, Pelegrin, F, Fernandez, J E and Murillo, J M (1997) 'Soil physical properties, water depletion and crop development under traditional and conservation tillage in southern Spain', *Soil & Tillage Research* 41(1–2), 25–42

Morin, J, Rowitz, E, Benyamini, Y, Hoogmoed, W B and Etkin, H (1984) 'Tillage practices for soil and water conservation in the semi arid zone. 2: Development of the basin-tillage system in wheat fields', *Soil & Tillage Research* 4(2), 155–164

Morin, J, Rowitz, E, Hoogmoed, W B and Benyamini, Y (1984) 'Tillage practices for soil and water conservation in the semi arid zone 3: Runoff modelling as a tool for conservation tillage design', *Soil & Tillage Research* 4(3), 215–224

Morton, J and Van Hoeflahen, H (1994) *Some Findings from a Survey of Spate Irrigation Schemes in Baluchistan, Pakistan* (ODI Irrigation Management Network Paper No 31). Overseas Development Institute, London

Mou Haisheng (1995) 'Rainwater utilization for sustainable development in North China', *Waterlines* 14(2), 19–21

Moussa, I B (1997) 'Valorisation des techniques anti-erosives traditionnelles en mileu sahelien: les micro-barrages en sacs de sable (Tuhoua, Niger) ', *Revue de Geographie Alpine* 85(1), 87–97

Moyersons, J (1994) 'Les essais recents de lutte anti-erosive en Rwanda', *Cahiers d'Outre-Mer* 47(185), 65–78

Murray, G F (1979) *Terraces, Trees, and the Haitian Peasant: an assessment of 25 years of erosion control in rural Haiti*. USAID/Haiti, AID–521–C–99 US Agency for International Development, Washington

Murray-Rust, D H and Rao, P S (1987) *Learning from Rehabilitation Projects: the Case of the Tank Irrigation Modernization Project (TIMP) of Sri Lanka* (ODI/IIMI Irrigation Management Network Papers No 87/2b). Overseas Development Institute, London

Nabhan, G P (1979) 'Ecology of floodwater farming in arid south-western North America', *Agro-Ecosystems* 5, 245–255

Nabhan, G P (1983) Papago fields: arid lands ethnobotany and agricultural ecology, PhD thesis. University of Arizona, Tucson

Nabhan, G P (1984) 'Soil fertility renewal and water harvesting in Sonoran Desert agriculture: the Papago example', *Arid Lands Newsletter* 20(1), 21–28

Nabhan, G P (1986a) 'Papago Indian desert agriculture and water control in the Sonoran Desert, 1697–1934', *Applied Geography* 6(1), 45–59

Nabhan, G P (1986b) '"AK-ciñ" "arroyo mouth" and the environmental setting of the Papago Indian fields in the Sonoran Desert' *Applied Geography* 6(1), 61–75

DI/IINational Academy of Sciences (1974) *More Water for Arid Lands: promising technologies and research opportunities* (Report of an Ad Hoc Panel of the Advisory Committee on Technology Innovation). NAS, Washington. (French and Spanish editions available)

Ndiaye, S M and Sofranko, A J (1994) 'Farmers' perceptions of resource problems and adoption of conservation measures', *Agricultural Ecosystems and Environment* 48(1), 35–47

Nessler, U (1980) 'Ancient techniques aid modern arid zone agriculture', *Kidma* 20(5), 3–7

Nevo, Y D (1991) *Pagans and Herders: a re-examination of the Negev runoff cultivation systems in the Byzantine and Early Arab periods*. IPS Ltd, Midrishet Ben-Gurion

Niemeijer, D (1998) 'Soil nutrient harvesting in indigenous *terras* water harvesting in Eastern Sudan', *Land Degradation & Development* 9(4), 323–330

Nilssen-Petersen, E (1982) *Rain Catchment and Water Supply in Rural Africa: a manual*. Hodder and Stoughton, London

Nilsson, Å (1988) *Groundwater Dams: for small-scale water supply*. Intermediate Technology Publications, London

Nye, P H and Greenland, D J (1960) *The Soil Under Shifting Cultivation* (CAB Technical Communication No 51). Commonwealth Agricultural Bureau, Harpenden

Odemerho, F and Avwunudiogba, A (1993) 'The effects of changing cassava management practices on soil loss: a Nigerian example', *The Geographical Journal* 159(1), 63–69

Ohlsson, L (ed) (1995) *Hydropolitics: conflicts over water as a development constraint*. Zed Books, London

Omoro, L M A and Nair, P V R (1993) 'Effects of mulching with multipurpose tree prunings on soil and water run-off under semi-arid conditions in Kenya', *Agroforestry Systems* 22(3), 225–239

Omotosho, J B (1993) 'Long-range prediction of the onset and end of the rainy season in the West African Sahel', *International Journal of Climatology* 12(4), 369–382

Oron, G, Heaton, P and Ben-Asher, J (1989) 'Design criteria for microcatchment water harvesting with scarce data', in J R Rydzewski and C F Ward (eds) *Irrigation Theory and Practice* (Proceedings of a Conference, Southampton, 1989). Wiley, Chichester, 302–316

Ortloff, C R (1988) 'Canal builders of pre-Inca Peru', *Scientific American* 259(6), 74–80

Oweis, T Y and Taimeh, A Y (1996) 'Evaluation of a small basin water harvesting system in the arid region of Jordan', *Water Resources Management* 10(1), 21–34

Owen, R (1995) *Dambo Farming in Zimbabwe: water management, cropping and soil potentials for smallholder farming in the wetlands* (Program, Recommendations and Papers from the Cornell International Institute for Food sponsored Conference, University of Zimbabwe, Harare, 8–10 September 1992). Cornell International Institute for Food, Agriculture and Development (CIIFAD) Cornell University, Ithaca

Pacey, A and Cullis, A (1986) *Rainwater Harvesting: the collection of rainfall and runoff in rural areas*. Intermediate Technology Publications, London

Pacey, A and Cullis, A (1992) *A Development Dialogue: rainwater harvesting in Turkana*. Intermediate Technology Publications, London

Palanisami, K (1990) *Tank Irrigation in South India: what next?* (ODI-IIMI Irrigation Management Network Paper No 9/2e). Overseas Development Institute, London

Palanisami, K and Easter, K W (1987) 'Small-scale surface (tank) irrigation in Asia', *Water Resources Research* 23(5), 774–780

Pandy, S (1991) 'The economics of water harvesting and supplementary irrigation in the semiarid tropics of India', *Agricultural Systems* 36(2), 207–220

Pangare, G (1992) 'Traditional water harvesting on way out', *Economic and Political Weekly* 17(10–11), A505

Park, C C (1983) 'Water resources and irrigation agriculture in pre-Hispanic Peru', *The Geographical Journal* 149(2), 153–166

Parry, M (1990) *Climate Change and World Agriculture*. Earthscan, London

Partap, T (1992) *The Last Oasis: facing water scarcity*. Earthscan, London

Partap, T and Watson, H R (1994) *Sloping Agricultural Land Technology (SALT). A Regenerative Option for Sustainable Mountain Farming* (ICIMOD Occasional Paper No 23). International Centre for Integrated Mountain Development, GPO Box 3226, Kathmandu, Nepal

Paul, B K (1984) 'Perception of an agricultural adjustment to floods in Jamuna flood-plain, Bangladesh', *Human Ecology* 12(1), 3–19

Payne, W A, Wendt, C W and Lasco, R J (1990) 'Bare fallowing on sandy fields of Niger, West Africa', *Soil Science Society of America Journal* 54(4), 1079–1084

Pearce, F (1992) *The Dammed: rivers, dams, and the coming world water crisis*. Bodley Head, London

Pearce, F (1997) 'Deluge of criticism greets Nile irrigation plan', *New Scientist* 153(2066), 5

Pellek, R (1992) 'Contour hedgerows and other soil conservation interventions for hilly terrain', *Agroforestry Systems* 17(2), 135–152

Pereira, H C (1989) *Policy and Practice in the Management of Tropical Watersheds*. Westview Press, Boulder

Pereira, L S, Gilley, J R and Jensen, M E (1996) 'Research agenda on sustainability of irrigated agriculture, ASCE', *Journal of Irrigation and Drainage Engineering* 122(3), 172–177

Pimentel, D (ed) (1993) *World Soil Erosion and Conservation*. Cambridge University Press, Cambridge

Pimentel, D, Harvey, C, Resosudarmo, P, Sinclair, K, McNair, M, Crist, S, Sgpritz, L, Fitton, L, Saffouri,R and Blair, R (1995) 'Environmental and economic costs of soil erosion and conservation benefits', *Science* 267(5201), 1117–1123

Pinche-Laurre, C (1996) 'Captacion de agua de niebla en lomas del la costa peruana', *Ingenieria Hidraulica en Mexico* 11(2), 49–54

Porter, G and Phillips-Howard, K (1997) 'Contract farming in South Africa: a case study from Kwa Zulu Natal', *Geography* 82(1), 38–44

Postel, S (1985) 'When rain only will do', *Development Forum* XII(9), 1; 4

Postel, S (1992) *The Last Oasis: facing water scarcity*. Earthscan, London

Pretty, J N and Shah, P (1997) 'Making soil and water conservation sustainable: from coercion and control to partnerships and participation', *Land Degradation & Development* 8(1), 39–58

Pretty, J N, Thompson, J and Kiara, J K (1995) 'Agricultural regeneration in Kenya: the catchment approach to soil and water conservation', *Ambio* XXIV(1), 7–15

Radford, B J, Key, A J, Robertson, L N and Thomas, G A (1995) 'Conservation tillage increases soil water storage, soil animal populations, grain yield, and response to fertilizer in the semiarid subtropics', *Australian Journal of Experimental Agriculture* 35(2), 223–232

Rahman, M (1981) 'The ecology of karez irrigation: a case of Pakistan', *GeoJournal* 5(1), 7–15

Rajaram, G, Erbach, D C and Warren, D M (1991) 'The role of indigenous tillage systems in sustainable food production', *Agriculture and Human Values* 8(1–2), 149–155

Rajeev, B M T (1992) 'A model treatment for gullies and ravines in Kalyanakere and Mavathurkere Watershed Development Programme', *MyForest* 28(1), 35–38

Ramírez, J A and Finnarty, B (1996) 'CO_2 and temperature effects on evapotranspiration and irrigated agriculture', *ASCE Journal of Irrigation and Drainage Engineering* 122(3), 155–163

Rao, V M and Chandrakant, M G (1984) 'Resources at the margin: tank irrigation programme in Karnataka', *Economic and Political Weekly* 19(26), A54–A58

Rapp, A and Hasteen-Dahlin, A (1990) 'The improved management of drylands by water harvesting in third world countries', in J Boardman, I D L Foster and J A Dearing (eds) *Soil Erosion on Agricultural Land (BGRG Symposia Series)*. Wiley, Chichester, 495–511

Raunet, M (1984) 'Les potentialités agricoles des bas-fonds en régions intertropicales: l'example de la culture du blé de contra-saison à Madagascar', *L'Agronomie Tropicale* 39(3), 181–201

Raunet, M (1985a) 'Bas-fonds et riziculture en Afrique: approche structurale et comparative', *L'Agronomie Tropicale* 40(2), 121–235

Raunet, M (1985b) 'Le bas-fonds en Afrique et la Madagascar: geomorphologie, geochemie, pedologie, hydrologie', *Zeitschrift für Geomorphologie supplementband* 52, 25–62

Reddy, D N, Barah, B C and Sudhakar, T (1993) 'Decline of traditional water harvesting systems in the drought prone areas of Andhra Pradesh', *Indian Journal of Agricultural Economics* 48(1), 76–87

Rees, D J, Qureshi, Z A, Mehmood, S and Raza, S H (1991) 'Catchment basin water harvesting as a means of improving the productivity of rain-fed land in upland Balochistan', *Journal of Agricultural Science* 116(1), 95–103

Reeve, I J and Black, A W (1994) 'Understanding farmers' attitudes to land degradation: some methodological considerations', *Land Degradation & Rehabilitation* 5(3), 179–190

Reij, C (1988) 'Soil and water conservation in sub-Saharan Africa: a bottom-up approach', *Appropriate Technology* 14(4), 14–16

Reij, C (1991) *Indigenous Soil and Water Conservation in Africa* (IIED Gatekeeper Series No 27). International Institute for Environment and Development, London

Reij, C, Mulder, P and Begemann, L (1988) *Water Harvesting for Plant Production* (World Bank Technical Paper No 91). World Bank, Washington

Reij, C, Scoones, I and Toulmin, C (eds) (1996) *Sustaining the Soil: indigenous soil and water conservation in Africa*. Earthscan, London

Reij, C, Turner, S and Kuhlman, T (1986) *Soil and Water Conservation in Sub-Saharan Africa: issues and options*. International Fund for Agricultural Development (IFAD), Africa Division, Rome

Reijntjes, C, Haverkort, B and Waters-Bayer, A (1992) *Farming for the Future: and introduction to low-external input and sustainable agriculture*. Macmillan, London

Renard, K G, Foster, G R, Weesies, G A and Porter, J F (1991) 'RUSLE – revised universal soil loss equation', *Journal of Soil & Water Conservation* 46(1), 30–33

Renner, H F and Frasier, G (1995a) 'Microcatchment water harvesting for agricultural production: Part I Physical and technical considerations', *Rangelands* 17(3), 72–78

Renner, H F and Frasier, G (1995b) 'Microcatchment water harvesting for agricultural production: Part II Socioeconomic considerations', *Rangelands* 17(3), 79–82

Richards, P (1985) *Indigenous Agricultural Revolution: ecology and food production in West Africa*. Hutchinson, London

Robineau, B and Robineau, M (1988) 'And the desert shall bloom again', *Waterlines* 7(2), 6–8

Rodriguez A, J and Lasanta Martinez, T (1992) 'Los bancales en la agricultura de la montana mediterranea: una reviso', *Pirineos* 139, 105–123

Rodriguez, A (1996) 'Sustainability and economic viability of cereals grown under alternative treatments of water harvesting in highland Balochistan, Pakistan', *Journal of Sustainable Agriculture* 8(1), 47–59

Rodriguez, A, Shah, N A, Afzal, M Mustafa, U and Ali,I (1996) 'Is water harvesting in valley floors a viable option for increasing cereal production in highland Balochistan, Pakistan', *Experimental Agriculture* 32(3), 305–315

Roggeri, H (1995) *Tropical Freshwater Wetlands: a guide to current knowledge and sustainable management*. Kluwer Academic, Dordrecht

Rohn, A H (1963) 'Prehistoric soil and water conservation on Chapin Mesa, southwestern Colorado', *American Antiquity* 28, 441–445

Rwejuna, C S (1994) 'Conservation strategies in Shinyanga: a need for mass mobilization', *Splash* 10(1), 13–14; 17–18

Sachs, C (1996) *Gendered Fields*. Westview Press, Boulder

Sanchez Cohen, I, Lopes, K L, Slack, D C and Fogel, M M (1997) 'Water balance model for small scale water harvesting systems', *ASCE Journal of Irrigation and Drainage Engineering* 123(2), 123–128

Sandar, J A, Gersper, P L and Hawley, J W (1990) 'Prehistoric agricultural terraces and soils in the Mimhes area, New Mexico', *World Archaeology* 22(1), 70–86

Sandys-Winsch, C and Harris, P J C (1994) '"Green" development on the Cape Verde Islands', *Environmental Conservation* 21(3), 225–230

Sauer, L (1991) 'Soil and water conservation in landscape perspective', *Journal of Soil & Water Conservation* 46(3), 194–196

Schemenauer, R S and Cereceda, P (1994) 'Fog collection's role in water planning for developing countries', *Natural Resources Forum* 18(2), 91–100

Schemenauer, R S and Cereceda, P (1997) 'Fog collection', *Tiempo* 26, 17–21

Schwab, G O and Frevert, R K (1981) *Soil and Water Conservation Engineering* (3rd edn) (1966 edn edited by R K Frevert). Wiley, New York

Scrimgeour, F G and Frasier, G W (1991) 'Runoff impoundment for supplemental irrigation: an economic assessment', *American Journal of Alternative Agriculture* 6(3), 139–145

Seitz, W D (1984) 'Who should pay how much for soil and water conservation? ', *Journal of Soil & Water Conservation* 39(5), 308–309

Shafiq, M, Hassan, A, Ahmad, S and Akhtar, M S (1994) 'Water intake as influenced by tillage in rainfed areas of the Punjab (Pakistan) ', *Journal of Soil & Water Conservation* 49(3), 302–305

Shakari, U (1991) 'Tanks: major problems in minor irrigation', *Economic & Political Weekly* 26(39), A115-A119

Shanan, L and Tadmore, N H (1979) *Micro-Catchment systems for Arid Zone Development. A Handbook for Design and Construction* (1st edn). Ministry of Agriculture, Centre for International Agricultural Cooperation, Rehovot (Israel)

Shanan, L, Evenari, M and Tadmore, N H (1969) 'Ancient technology and modern science applied to desert agriculture', *Endeavour* 28(1), 66–73

Shankari, U (1991) 'Major problems in minor irrigation: social change and tank irrigation in Chittoor district of Andhra Pradesh', *Contributions to Indian Sociology* 25(1), 85–111

Sharma, K D (1986) 'Runoff behavior of water harvesting microcatchments', *Agricultural Water Management* 11(2), 137–144

Sharma, K D, Pareek, O P and Singh, H P (1986) 'Microcatchment water harvesting for raising jujube orchards in an arid climate', *Transactions of the ASAE* 29(1), 112–118

Sharma, K D, Vangani, N S, Singh, H P, Bohra, D N, Kella, A K and Joshi, P K (1997) 'Evaluation of contour vegetative barriers as soil and water conservation measures in arid India', *Annals of Arid Zone* 36(2), 123-127

Shaxson, T F (1997) *Better Land Husbandry: re-thinking approaches to land improvement and the conservation of water and soil* (ODI Natural Resource Perspectives No 19). Overseas Development Institute, London

Shaxson, T F, Hudson, N W, Sanders, D W, Roose, E and Moldenhauer, W C (1989) *Land Husbandry: a framework for soil and water conservation.* Soil and Water Conservation Society, Ankeny

Shemenauer, R S and Cereceda, P (1991) 'Fog collection in arid coastal locations', *Ambio* XX(7), 303–308

Shemenauer, R S and Cereceda, P (1992) 'Water from fog-covered mountains', *Waterlines* 10(4), 10–13

Shemenauer, R S and Cereceda, P (1994) 'Fog collection's role in water planning for developing countries', *Natural Resources Forum* 18(2), 91–100

Sheng, T C (1989) *Soil Conservation for Small Farmers in the Humid Tropics.* FAO, Rome

Siegert, K (1995) 'Gathering the rains', *Planter* 71(837), 591–595

Sillitoe, P (1993) 'Losing ground? Soil loss and erosion in the highlands of Papua New Guinea', *Land Degradation & Rehabilitation* 4(3), 143–166

Singh, G (1976) 'Watershed management administrations', *Unasylva* 28(114), 32–38

Singh, P K, Modi, S Mahnot, S C and Singh, J (1995) 'Watershed approach in improving the socio-economic status of tribal areas: a case study', *Journal of Rural Development* 14(2), 107–116

Singh, S D (1985) 'Potentials of water harvesting in the dry regions', *Annals of Arid Zone* 24(1), 9–20

Skarie, R and Bloom, P (1982) *Maya Subsistence.* Academic Press, New York

Sluyter, A (1994) 'Intensive wetland agriculture in MesoAmerica: space, time and form', *Annals of the Association of American Geographers* 84(4), 557–584

Smit, B and Smithers, J (1992) 'Adoption of soil conservation practices: an empirical analysis in Ontario, Canada', *Land Degradation & Rehabilitation* 3(1), 1–14

Smith, M E and Price, T J (1994) 'Aztec period agricultural terraces in Morelos, Mexico', *Journal of Field Archaeology* 21(2), 169–179

Soil and Tilage Research (1993) 'Special Issue. Proceedings of the 12th Conference of Istro Part II. Soil tillage for agricultural sustainability', 27(1–4), 1–385

Sombatpanit, S, Zöbisch, M A, Sanders, D W and Cook, M G (eds) (1997) *Soil Conservation Extension: from concepts to adoption.* Science Publishers Inc, Enfield (New Hampshire, USA)

Sorman, A U, Abdulrazzak, M J and Elhames, A S (1990) 'Rainfall-runoff modelling of a microcatchment in the western region of Saudi Arabia', in H Lang and A Musy (eds) *Hydrology in Mountainous Regions: I* (IAHS/AISH Publications No 193). International Association of Hydrological Sciences, Wallingford, 655–659

Spoor, G and Berry, R H (1990) 'Dryland farming, tillage and water harvesting guidelines for the Yemen Arab Republic', *Soil & Tillage Research* 16(1–2), 233–244

Srivastava, J P, Tamboli, P M, English, J C, Lal, R and Stewart, B A (1993) *Conserving Soil Moisture and Fertility in the Warm Seasonally Dry Tropics* (World Bank Technical Paper No 21). World Bank, Washington

Stern, P (1979) *Small Scale Irrigation: a manual of low-cost water technology.* Intermediate Technology Publications, London

Stern, P (1988) *Operation and Maintenance of Small Irrigation Schemes.* Intermediate Technology Publications, London

Stiles, D (ed) (1995) *Social Aspects of Sustainable Dryland Management.* Wiley, Chichester

Stocking, M (1988) 'Socioeconomics of soil conservation in developing countries', *Journal of Soil & Water Conservation* 43(5), 381–385

Stone, E L (1957) 'Dew as an ecological factor. A review of the literature', *Ecology* 38(3), 407–417

Stonehouse, D P (1995) 'Profitability of soil and water conservation in Canada: a review', *Journal of Soil & Water Conservation* 50(2), 215–219

Suleman, S, Wood, M K, Shah, B H and Murray, L (1995) 'Development of a rainwater harvesting system for increasing soil moisture in arid rangelands of Pakistan', *Journal of Arid Environments* 31(4), 471–481

Sur, H S, Mastana, P S and Hadda, M S (1992) 'Effect of rates and modes of mulch application on runoff, sediment and nitrogen loss on cropped and uncropped fields', *Tropical Agriculture* 69(4), 319–322

Sutherland, R A (1998) 'Rolled erosion control systems for hillslope surface protection: a critical review, synthesis and analysis of available data'. I Background and formative years, *Land Degradation & Development* 9(6), 465–486

Sutherland, R A and Bryan, R B (1990) 'Runoff and erosion from a small catchment, Baringo Catchment, Kenya', *Applied Geography* 10(2), 91–109

Sutton, J E G (1984) 'Irrigation and soil conservation in African agricultural history: with a reconsideration of the Inyanga terracing (Zimbabwe) and Engaruka irrigation works (Tanzania)', *Journal of African History* 25(1), 25–41

Sutton, S (1984) 'The Falaj: a traditional co-operative system of water management (in Oman)', *Waterlines* 2(3), 8–12

Swanson, L E, Camboni, S M and Napier, T L (1986) 'Barriers to adoption of soil conservation practices on farms', in S B Lovejoy and T L Napier (eds) *Conserving Soil: insights from socio-economic research.* Soil Conservation Society of America, Ankeny, 108–120

Taabni, M and Kouti, A (1993) 'Strategies de conservation, mise en oeuvre et reactions du milieu et des paysans dans l'ouest algerien', Association de Geographes Français 1993–1995, 408–422

Tabor, J A (1995) 'Improving crop yields in the Sahel by means of water-harvesting', *Journal of Arid Environments* 30(1), 83–106

Tagwira, F (1992) 'Soil erosion and conservation techniques for sustainable crop production in Zimbabwe', *Journal of Soil & Water Conservation* 47(5), 370–374

Taller, W, Prinz, D and Vogtle, T (1991) 'The potential of runoff-farming in the Sahel region: developing a methodology to identify suitable areas', *Water Resources Management* 5(3–4), 281–287

Tanaka, D L and Anderson, R L (1997) 'Soil water storage and precipitation storage efficiency of conservation tillage systems', *Journal of Soil & Water Conservation* 52(5), 363–367

Tato, K and Hurni, H (eds)(1992) *Soil Conservation for Survival* (Soil Conservation Research Report). Centre for Development and Environment, Institute of Geography, University of Berne, Berne

Tauer, W and Humborg, G (1992) *Runoff Irrigation in the Sahel Zone: remote sensing and geographical information systems for determining potential sites* (Technical Centre for Agricultural and Rural Co-operation, Agricultural University, Wageningen). Verlag Josef Margraf, Wageningen

Tauer, W, Prinz, D, Vogtle, T (1991) 'The potential of runoff-farming in the Sahel region: developing a methodology to identify suitable areas', *Water Resources Management* 5(3–4), 281–287

Tenbergen, B, Günster, A and Schreiber, K–F (1995) 'Harvesting runoff: the minicatchment technique – an alternative to irrigated tree plantations in semiarid regions', *Ambio* XXIV(2), 72–76

Ternan, J L, Williams, A G, Elmes, A and Fitzjohn, C (1996) 'The effectiveness of bench terracing and afforestation for erosion control on Rana sediments in central Spain', *Land Degradation & Development* 7(4), 337–351

Thapa, G B and Weber, K E (1995) 'Natural resource degradation in a small watershed in Nepal: complex causes and remedial measures', *Natural Resources Forum* 19(4), 285–296

The Times 29 November 1997, 30

Thomas-Slayter, B and Rocheleau, D (1995) *Gender, Environment and Development in Kenya: a grassroots perspective.* Lynne Rienner, Boulder

Thompson, D A and Scoging, H M (1995) 'Agricultural terrace degradation in southeast Spain: modelling and management', in D F M McGregor and D A Thompson (eds) *Geomorphology and Land Management in a Changing Environment.* Wiley, Chichester, pp 153–175

Thurow, T L and Juo, A S R (1995) 'The rationale for using a watershed as a basis for planning and development', in A S R Juo and R D Freed (eds) *Agriculture and Environment: bridging food production and environmental protection in developing countries* (ASA Special Publication No 60). American Society of Agronomy, Ankeny, 93–116

Tiffen, M (1995) 'Population density, economic growth and societies in transition: Boserüp reconsidered in a Kenyan case study', *Development & Change* 26(1), 32–65

Tiffen, M, Mortimore, M and Gichuki, F (1994) *More People, Less Erosion: environmental recovery in Kenya.* Wiley, Chichester

Treacy, J M (1989) 'Agricultural terraces in Peru's Colca Valley: promises and problems of an ancient technology', in J O Browder (ed) *Fragile Lands in Latin America: strategies for sustainable development.* Westview Press, Boulder, 209–229

Troeh, F R, Hobbs, J A and Donahue, R L (1980) *Soil and Water Conservation for Productivity and Environmental Protection* (with editorial assistance from M R Troeh). Prentice-Hall, Englewood Cliffs

Tsakiris, G (1991) 'Micro-catchment water harvesting in semi-arid regions: basic design considerations', *Water Resources Management* 5(1), 85–92

Tsioutis, N X (1995) *Water Resources Management Under Drought or Water Shortage Conditions.* A A Balkema, Rotterdam

Tuan, C H (1988) 'Study on the gully control by used tire structure in northern Taiwan', in C Bonnard (ed) *Landslides (Proceedings of the 5th Symposium on Landslides, Lausanne, 1988).* A A Balkema, Rotterdam

Turner, B (1994) 'Small scale irrigation in developing countries', *Land Use Policy* 11(4), 251–261

Turner, K and Jones, T (eds) (1991) *Wetlands: market and intervention failures.* Earthscan, London

Turton, C and Bottrall, A (1997) 'Water resource development in the drought-prone uplands', *ODI Natural Resource Perspectives* No 18. Overseas Development Institute, London

Uitto, J I and Schneider, J (eds)(1997) *Freshwater Resources in Arid Lands* (UNU Global Environmental Forum, vol V). United Nations University Press, Tokyo

Unasylva (1986) 152, 38

UNEP (1983) *Rain and Stormwater Harvesting in Rural Areas: a report by the UNEP.* Tycooly International, Dublin

UNESCO (1962) *The Problems of the Arid Zone: arid zone research* (vol XVIII). UNESCO, Paris

Unger, P W and Cassel, D K (1991) 'Tillage implement disturbance effects on soil properties related to soil and water: a literature review', *Soil & Tillage Research* 19(4), 363–382

eae= klac	ttel a

Uphoff, N (1986) *Improving International Irrigation Management with Farmer Participation: getting the process right*. Westview Press, Boulder

Valdes, J B, Seoane, R S and North, G R (1994) 'A methodology for evaluation of global warming impact on soil-moisture and runoff', *Journal of Hydrology* 161(1–4), 389–413

Vandermeer, J (1992) *The Ecology of Intercropping*. Cambridge University Press, Cambridge

Van der Ploeg, J D and Van Dijk, G (eds) (1995) *Beyond Modernization: the impact of endogenous rural development*. Van Gorcum Ltd, Assen (The Netherlands),

Van Der Waal, T and Zaal, F (1990) *Bibliography on Indigenous Soil and Water Conservation with Special Reference to Africa*. Vrei Universiteit Amsterdam, Centre for Development Cooperation Services, Amsterdam

Van Dijk, J A (1995) 'Indigenous and introduced soil and water conservation in Sudan', *Waterlines* 13(4), 19–21

Van Dijk, J A (1997a) 'Indigenous soil and water conservation by *teras* in eastern Sudan', *Land Degradation & Development* 8(1), 17–26

Van Dijk, J A (1997b) 'Simple methods for soil moisture assessment in water harvesting projects', *Journal of Arid Environments* 35(3), 387–394

Van Dijk, J A and Ahmed, M H (1993) *Opportunities for Expanding Water Harvesting in Sub-Saharan Africa: The Case of the Terras of Kassala* (Gatekeeper Series No SA40). International Institute of Environment and Development (IIED), London

Van Dijk, J A and Reij, C (1994) 'Indigenous water harvesting techniques in sub-Saharan Africa: examples from Sudan and the West African Sahel', in FAO (ed) *Water Harvesting for Improved Agricultural Production* (Proceedings of the FAO Expert Consultation – Water Reports No 3). FAO, Rome, 101–112

Van Immerzeel, W and Osterbaan, R J (1990) 'Irrigation and flood erosion control at high altitudes in the Andes', in ILRI (ed) *ILRI Annual Report 1989*. International Land Reclamation Institute, Wageningen, 8–24

Van Steenbergen, F (1997) 'Understanding the sociology of spate irrigation: cases from Balochistan', *Journal of Arid Environments* 35(2), 349–365

Varisco, D M (1983) 'Irrigation in an Arabian valley: systems of highland terraces in the Yemen Arab Republic', *Expedition* 25(2), 26–34

Varisco, D M (1991) 'The future of terrace farming in north Yemen: a development dilemma', *Agriculture and Human Values* 8(1–2), 166–172

Veek, G, Li Zhou and Ling, G (1995) 'Terrace construction and productivity on loessal soils in Zhongyang County, Shanxi Province, PRC', *Annals of the Association of American Geographers* 85(3), 450–467

Verkruysse, B (1992) *Annotated Bibliography: gender and irrigation and soil and water conservation*. Department of Gender Studies in Agriculture, State Agricultural University, Wageningen

Verma, H N and Sarma, P B S (1990) 'Design of storage tanks for water harvesting in rainfed areas', *Agricultural Water Management* 18(3), 195–207

Victor, U S, Srivastava, N N and Rao, B V R (1991) 'Application of crop water use models and rainfed conditions in seasonally arid tropics', *International Journal of Ecology and Environmental Sciences* 17(2), 115–118

Vincent, L (1990a) 'Sustainable small scale irrigation development: issues for farmers, governments and donors', *International Journal of Water Resources Development* 6(4), 250–259

Vincent, L (1990b) 'Environmentally sound irrigation projects', *Waterlines* 8(4), 3–6

Vincent, L (1995) *Hill Irrigation Water and Development in Modern Agriculture*. Intermediate Technology Publications, London

Vlaar, J C J (1992) 'Design and efficiency of permeable infiltration dams in Burkina Faso', *Land Degradation & Rehabilitation* 3(1), 37–53

Vogel, H (1985) 'Terrace farming in the Yemen Arab Republic Traditional forms of soil and water conservation and their present degradation, a case study of the Manakhah Region', Paper presented to the Fourth International Conference on Soil Conservation, 3–9 November, 1985. Maracay, Venezuela

Vogel, H (1992) 'Effects of conservation tillage on sheet erosion from sandy soils at two experimental sites in Zimbabwe', *Applied Geography* 12(3), 229–242

Vogel, H (1993) *Soil erosion in terrace farming: destruction of farming land induced by changes in land use in the Hraz Mountains of Yemen* (Deutsche Gesellschaft für Technische Zusammenarbeit). GTZ, Eschborn

Vogen, H (1988) 'Deterioration of a mountain agro-ecosystem in the third world due to emigration of labour', *Mountain Research & Development* 8(4), 321–329

Von Carlowitz, P G and Wolf, G V (1991) 'Open-pit planting: a tree establishment technique for dry environments', *Agroforestry Systems* 15(1), 17–29

Von Oppen, M and Subba Rao, K V (1987) *Tank Irrigation in Semi-Arid Tropical India: economic evaluation and alternatives for improvement* (ICRISAT Research Bulletin No 10). International Centre for Research in the Semi-Arid Tropics (ICRISAT), Andhra Pradesh

Walter, H (1971) *Ecology of Tropical and Sub-Tropical Vegetation* (2nd edn). Oliver and Boyd, Edinburgh

Wardman, A and Salas, L G (1991) 'The implementation of anti-erosion techniques in the Sahel: a case study from Kaya, Burkina Faso', *Journal of Developing Areas* 26(1), 65–80

Warren, D M (1991) *Using Indigenous Knowledge in Agricultural Development* (World Bank Discussion Paper No 127). The World Bank, Washington

Warren, D M, Fujisaka, S and Prain, G (eds) (1998) *Biological and Cultural Diversity: the role of indigenous agricultural experimentation in development*. Intermediate Technology Publications, London

Water, Engineering and Development Centre (WEDC) (undated) 'Technical Brief No 24 – Groundwater Dams', *Waterlines* 8(4), 15–18

Webster, C C and Wilson, P N (1966) *Agriculture in the Tropics* (1st edn). Longman, Harlow

Weischmeir, W H and Smith, D D (1960) 'A universal soil loss estimating equation to guide conservation farm planning', *Transactions of the 7th International Congress of Soil Science* 1, 418–425

Weischmeir, W H and Smith, D D (1978) *Predicting Rainfall Erosion Losses* (USDA Handbook No 537). US Department of Agriculture, Washington

Wheatley, P (1965) 'Agricultural terracing', *Pacific Viewpoint* 5–6, 123–144

White, T A and Runge, C F (1994) 'Common property and collective action: lessons from cooperative watershed management in Haiti', *Economic Development & Cultural Change* 43(1), 1–41

White, T A and Runge, C F (1995a) 'The emergence and evolution of collective action: lessons from a watershed management in Haiti', *World Development* 23(10), 1683–1698

White, T A and Runge, C F (1995b) 'Cooperative watershed management in Haiti: common property and collective action', *Unasylva* 180, 50–37

Wiese, A F, Harman, W L, Bean, B W and Salisbury, C D (1994) 'Effectiveness and economics of dryland conservation tillage systems in the southern Great Plains', *Agronomy Journal* 86(4), 725–730

Wilken, C G (1987) *Good Farmers: traditional agricultural resource management in Mexico and Central America*. University of California Press, Berkeley

Wilkinson, J C (1974) *The Organisation of the Falaj Irrigation System in Oman* (University of Oxford, School of Geography Research Papers No 10). University of Oxford, School of Geography, Oxford

Wong, S J (1984) 'The effectiveness of conservation farms used by the Soil Conservation Service and soil and water conservation districts: a dual perspective', *Ohio Journal of Science* 84(2), 47

World Bank (1988) *Vetiver grass: a method of vegetative soil and water conservation* (2nd edn). World Bank, New Delhi

Wrigley, G (1981) *Tropical Agriculture: the development of production* (4th edn). Longman, London

Wulff, H E (1969) 'Quanats of Iran', *Scientific American* 218(4), 94–105

Yadav, R C, Kumar, V and Oguntela, V B (1983) 'Biological water harvesting: a method of enabling dryland crops to endure periods of drought', *Journal of Arid Environments* 6(2), 115–118

Yair, A (1983) 'Hillslope hydrological water harvesting and areal distribution of some ancient agricultural systems in the northern Negev Desert', *Journal of Arid Environments* 6(3), 283–302

Yoder, R (ed)(1994) *Designing Irrigation Structures for Mountainous Environments: a handbook of experience*. International Irrigation Management Institute, Colombo

Young, A (1989) *Agroforestry for Soil and Water Conservation* (ICRAF Science and Practice of Agroforestry No 4). CAB International, Wallingford

Zanen, Sj M (1995) 'Sustainable bas-fond development in Burkina Faso', in H Roggeri (ed) *Tropical Freshwater Wetlands: a guide to current knowledge and sustainable management*. Kluwer Academic, Dordrecht, 167–172

Zimmerer, K S (1994) 'Soil erosion and labour shortages in the Andes with special reference to Bolivia, 1953–91: implications for "conservation-with-development"' *World Development* 21(10), 1659–1675

Zingg, A W and Hauser, V L (1959) 'Terrace benching to save potential runoff for semiarid land', *Agronomy Journal* 51(3), 289–292

Zinn, J A (1988) 'Cataloging experiences: the USDA approach. Reporting progress – the USDA approach to soil and water conservation (editorial) ', *Journal of Soil & Water Conservation* 43(2), 152

Zurayk, R A (1994) 'Rehabilitating the ancient terraced lands of Lebanon', *Journal of Soil & Water Conservation* 49(2), 106–112

Zwarteveen, M Z (1997) 'Water: from basic needs to commodity. A discussion on gender and water rights in the context of irrigation', *World Development* 25(8), 1351–1371

FURTHER READING

Some journals and abstracts which publish articles on SWC and runoff agriculture:

Advances in Agronomy
Agricultural Ecosystems and Environment
Agricultural Water Management
Agroforestry
Agroforestry Systems
Agronomy Abstracts
Annals of Arid Zone ·
Arid Soil Research & Rehabilitation
ASCE Journal of Irrigation and Drainage Engineering
Australian Journal of Experimental Agriculture
CAB Abstracts
Conservation Farming
International Development Abstracts
International Journal of Tropical Agriculture
IRRI News (Volcani Center, PO Box 49, Bet Dagan, Israel IL–50250)
Irrigation and Drainage Systems
Journal of Arid Environments
Journal of Hydrology
Journal of Soil & Water Conservation
Journal of Sustainable Agriculture
Land Degradation & Development
Land Degradation & Society
Restoration Ecology
Society & Natural Resources
Soil Conservation
Soil & Tillage Research
Soil and Water
Soil Technology
Soil Use and Management
Tropical Agriculture
Water Resources Management
Waterlines

INDEX